Anonymous

Charles Millar and Son, Utica N.Y., U.S.A.

Wrought iron pipe, cast iron pipe, lead pipe, block tin pipe, akron vitrified sewer pipe ; valves fittings and supplies of every description for water works, plumbers, steam and gas fitters

Anonymous

Charles Millar and Son, Utica N.Y., U.S.A.
Wrought iron pipe, cast iron pipe, lead pipe, block tin pipe, akron vitrified sewer pipe ; valves fittings and supplies of every description for water works, plumbers, steam and gas fitters

ISBN/EAN: 9783337377076

Printed in Europe, USA, Canada, Australia, Japan

Cover: Foto ©berggeist007 / pixelio.de

UTICA, N. Y.:
PRESS OF L. C. CHILDS & SON.
1894.

TO THE TRADE.

IN ISSUING OUR NEW CATALOGUE of Engineers', Plumbers', Gas and Steam Fitters' Supplies, we wish to express our grateful appreciation of the generous patronage extended to our house for over thirty years.

With improved and increased facilities and an earnest desire to faithfully, promptly and carefully serve our friends and patrons, we hope to command a larger share of their trade than ever before.

We are largely interested in and are the exclusive Selling Agents for the Utica Pipe Foundry Co., and are prepared to furnish all material for the construction of **new** water works or the extension and improvement of old ones. Our Pipe will be found of superior quality and workmanship and made from higher grades of iron than other makes.

Our reputation for making the best Lead Pipe in the country has been and will be maintained and is fully demonstrated by increased sales; and the same is true of our Solder, Tin Pipe and Tubing.

We continue to act as agents for various manufacturers of patented and staple goods in our line, have large warehouses with full stocks, and are in position to offer unequalled advantages to dealers.

Soliciting your valued trade and inviting correspondence, we are

Very respectfully,

CHARLES MILLAR & SON.

CONDENSED INDEX.

INDEX.

PRICE LIST

OF

Wrought Iron Pipe and Fittings, for Steam, Gas or Water.

Standard Wrought-Iron Steam, Gas and Water Pipe.

LIST ADOPTED APRIL 13, 1893.

BUTT-WELDED.

Nominal Size Inside Diameter.	Price per Foot, Black.	Price per Foot. Galvanized or Rustless.	Price per ft., Tarred or Asphalted.	Actual Outside Diameter.	Thickness.	Nominal Weight per Foot.	No. of Threads per Inch of Screw.
INCHES.				INCHES.	INS.	LBS.	
1/8	$ .04	$ .05	....	.40	.068	0.24	27
1/4	.04	.05	$ .04½	.54	.088	0.42	18
3/8	.04½	.05½	.05½	.67	.091	0.56	18
1/2	.06	.08	.07	.84	.109	0.84	14
3/4	.07½	.10	.10	1.05	.113	1.12	14
1	.11	.14	.12½	1.31	.134	1.67	11½
1¼	.14½	.19	.17	1.66	.140	2.24	11½

LAP-WELDED.

Nominal Size Inside Diameter.	Price per Foot, Black.	Price per Foot, Galvanized or Rustless.	Price per ft., Tarred or Asphalted.	Actual Outside Diameter.	Thickness	Nominal Weight per Foot.	No. of Threads per Inch of Screw.
INCHES.				INCHES.	INCHES.	POUNDS.	
1½	$ .24	$ .28	$ .30	1.9	.145	2.68	11½
2	.33	.38	.39	2.37	.154	3.61	11½
2½	.50	.57	.58	2.87	.204	5.74	8
3	.64	.70	.73	3.5	.217	7.54	8
3½	.76	.90	.91	4.	.226	9.00	8
4	.90	1.05	1.10	4.5	.237	10.66	8
4½	1.06	1.31	1.36	5.	.247	12.34	8
5	1.28	1.60	1.58	5.56	.259	14.50	8
6	1.65	2.00	2.00	6.62	.280	18.76	8
7	2.10	....	....	7.62	.301	23.27	8
8	2.75	....	....	8.62	.322	28.18	8
9	3.75	....	...	9.68	.344	33.70	8
10	4.75	....	....	10.75	.366	40.00	8
11	6.00	....	....	11.75	.375	45.00	8
12	7.00	....	...	12.75	.375	49.00	8
13	8.00	...	....	14.	.375	54.00	8
14	9.50	....	....	15.	.375	58.00	8
15	11.00	....	....	16.	.284	62.00	8

For Selected Pipe or Pipe cut to specified lengths, the discount will be Five (5) per cent. less in the gross than on Regular Pipe.

Extra Strong and Double Extra Strong Wrought-Iron Pipe.

BUTT-WELDED.

SIZE.	Price per Foot.	Actual Outside Diameter.	Nominal Inside Diameter.	Thickness.	Nominal Weight per Foot.
X Strong.					
1/8	$ .08	.40	.205	.100	.29
1/4	.08	.54	.294	.123	.54
3/8	.09	.67	.421	.127	.74
1/2	.12	.84	.542	.149	1.09
3/4	.15	1.05	.736	.157	1.39
1	.22	1.31	.951	.182	2.17
1¼	.29	1.66	1.272	.194	.
XX Strong.					
1/2	.24	.84	.244	.298	1.70
3/4	.30	1.05	.422	.314	2.44
1	.44	1.31	.587	.364	3.65
1¼	.58	1.66	.885	.388	5.20

LAP-WELDED.

SIZE.	Price per Foot.	Actual Outside Diameter.	Nominal Inside Diameter.	Thickness.	Nominal Weight per Foot.
X Strong.					
1½	$ .48	1.90	1.494	.203	3.63
2	.66	2.375	1.983	.221	5.02
2½	1.00	2.875	2.315	.280	7.67
3	1.28	3.50	2.892	.304	10.25
3½	1.52	4.00	3.358	.321	12.47
4	1.80	4.50	3.818	.341	14.97
5	2.56	5.563	4.813	.375	20.54
6	3.30	6.625	5.750	.437	28.58
XX Strong.					
1½	.96	1.90	1.088	.406	6.40
2	1.32	2.375	1.491	.442	9.02
2½	2.00	2.875	1.755	.560	13.68
3	2.56	3.50	2.284	.608	18.56
3½	3.04	4.00	2.716	.642	22.75
4	3.60	4.50	3.136	.682	27.48
5	5.12	5.563	4.063	.75	38.12
6	6.60	6.625	4.875	.875	53.11

SEAMLESS AND DRAWN BRASS AND COPPER TUBING. IRON PIPE SIZE.

Same Size as Iron Pipe.	BRASS.		COPPER.		Actual Outside Diameter. Inches.	Actual Inside Diameter. Inches.
	Extras Over Basis Price per lb.	Approximate Weight per Foot.	Extras over Basis Price per lb.	Approximate Weight per Foot.		
1/8	$0.15	.296	$0.15	.310	.405	.250
1/4	.11	.451	.11	.476	.540	.343
3/8	.08	.600	.08	.657	.675	.468
1/2	.06	.780	.06	.838	.840	.625
3/4	Basis	1.180	Basis	1.259	1.050	.843
1	"	1.750	"	1.855	1.315	1.062
1¼	"	2.535	"	2.681	1.660	1.343
1½	"	3.045	"	3.166	1.900	1.593
2	"	4.000	"	4.369	2.375	2.062
2½	"	6.323	"	6.868	2.875	2.437
3	"	8.500	"	8.979	3.500	3.060
3½	.02	9.878	.02	10.646	4.000	3.500
4	.02	11.719	.02	12.730	4.500	4.000
5	.06	16.935	.06	17.309	5.560	5.040
6	.10	21.199	.10	22.385	6.620	6.060
7	.15	26.286	.15	27.777	7.620	7.020
8	.20	29.881	.20	33.690	8.620	7.980

Elbows.

SIZE	1/8	1/4	3/8	1/2	3/4	1	1¼	1½	2	2½	3	3½	4	4½	5	6	7	8	10 in.
Iron	....	.04	.05	.06	.09	.13	.20	.25	.40	.75	1.10	1.35	1.80	2.50	2.85	3.90	7.00	10.00	20.00
" Reducing	....	.05	.06	.07	.11	.16	.23	.29	.46	.85	1.25	1.50	2.10	3.00	3.25	4.50	8.00	11.50	23.00
" Right and Left	....	.05	.06	.07	.11	.16	.23	.29	.46	.85	1.25	1.50	2.10				....	....	....
" with Side Outlet	....	.08	.10	.12	.18	.26	.40	.50	.80	1.50	2.20	2.70	3.60	5.00	5.70	7.80	....	....	....
Galvanized	....	.06	.09	.12	.18	.30	.45	.55	.85	1.60	2.35	3.10	4.10	6.00	7.00	11.00	....	....	....
" Reducing	....	.10	.12	.14	.22	.32	.46	.58	.92	1.70	2.50	3.00	4.20	6.00	....	....	....	....	....
Brass, Rough	.10	.12	.16	.25	.35	.50	.85	1.15	1.50	3.00	5.00	....	....	....	....	....	....	....	....
" Finished	.15	.18	.24	.35	.45	.65	1.10	1.50	2.00	3.75	6.50	....	....	....	....	....	....	....	....

Street and Drop Elbows Same Price as Reducing Elbows.

Union Elbows.

SIZE	1/2	3/4	1	1¼	1½	2 in.
Plain	.28	.35	.45	.65	.80	1.25

45° Elbows.

SIZE	1/8	1/4	3/8	1/2	3/4	1	1¼	1½	2	2½	3	3½	4	4½	5	6	7	8	10 in.
Iron	...	.08	.10	.10	.15	.20	.26	.35	.50	1.30	1.60	1.90	2.50	3.50	4.50	5.50	9.00	12.00	22.00
Galvanized	...	....	.20	.20	.30	.40	.50	.70	1.00	2.60	3.20	4.75	5.00	7.00	9.00	11.00	18.00	24.00	....
Brass, Rough	.15	.20	.25	.35	.50	.70	1.05	1.40	1.80	3.75	6.00	...	...	...	...	...	...	...	...
" Fin.	....	.30	.35	.50	.65	.85	1.25	1.75	2.30	4.50	7.50	...	...	...	...	...	...	...	...

Tees.

SIZE	1/8	1/4	3/8	1/2	3/4	1	1¼	1½	2	2½	3	3½	4	4½	5	6	7	8	10 in.
Iron	....	.06	.07	.09	.18	.20	.30	.38	.60	1.10	1.50	2.00	2.50	3.50	4.00	5.50	10.00	15.00	25.00
" Reducing	....	.07	.08	.11	.15	.23	.35	.44	.70	1.25	1.75	2.30	2.90	4.00	4.60	6.35	......	......	......
Galvanized	....	.08	.13	.17	.25	.40	.66	.85	1.20	2.25	2.85	3.80	5.25	7.00	8.00	12.50	......	......	......
" Reducing	....	.14	.16	.22	.30	.46	.70	.88	1.40	2.50	3.50	4.60	5.80	8.00	9.20	12.70	......	......	......
Brass, Rough	.12	.15	.20	.30	.45	.70	1.00	1.25	1.75	4.00	6.00	....	....	......	......	......	......	......	......
" Finished	.20	.25	.35	.50	.65	1.00	1.50	2.00	2.75	5.50	8.00	....	....	......	......	......	......	......	......

Union Tees.

SIZE	½	¾	1	1¼	1½	2 in.
Plain	.28	.55	.45	.65	.80	1.25

Crosses.

SIZE	1/8	1/4	3/8	1/2	3/4	1	1¼	1½	2	2½	3	3½	4	4½	5	6	7	8	10 in.
Iron	....	.08	.10	.12	.18	.28	.40	.50	.80	1.50	2.20	2.70	3.50	5.00	5.70	7.80	14.00	20.00	40.00
" Reducing	....	.10	.12	.14	.21	.32	.46	.58	.92	1.70	2.50	3.00	4.00	6.00	6.60	9.00	......	......	......
Galvanized	....	.15	.18	.23	.35	.55	.80	1.00	1.60	3.00	4.25	5.50	7.00	9.00	10.50	16.00	......	......	......
" Reducing	....	.20	.24	.28	.42	.64	.92	1.16	1.84	3.40	5.00	6.00	8.00	12.00	13.20	18.00	......	......	......
Brass, Rough	.15	.20	.30	.40	.55	.80	1.15	1.40	1.95	4.25	6.50	....	....	......	......	......	......	......	......
" Finished	.30	.35	.55	.70	.85	1.25	1.90	2.50	3.50	6.25	9.50	....	....	......	......	......	......	......	......

Return Bends.

SIZE	½	¾	1	1¼	1½	2	2½	3 in.
Iron, Close	.10	.15	.22	.34	.45	.75	1.50	2.25
" Open	.15	.20	.30	.48	.68	1.15	1.75	2.75
Galvanized	.17	.25	.40	.60	.85	1.50	....	......
Brass, Close	.40	.70	.90	1.25	2.00	2.75	6.00	9.00
" Open	.50	.80	1.00	1.50	2.25	3.00	7.50	11.00
Finished Brass, Open	.70	1.00	1.80	2.00	3.00	4.00	...	......

Offsets.

SIZE	¾	1	1¼	1½	2	2½	3	3½	4	5	6 in.
To Offset 4 Inches	.45	.70	1.00	1.20	1.80	3.00	4.00	5.00	6.00	8.00	10.00
" 6 "	.67	1.05	1.50	1.80	2.70	4.50	6.00	7.50	9.00	12.00	15.00
" 8 "	.90	1.40	2.00	2.40	3.60	6.00	8.00	10.00	12.00	16.00	20.00

Unions.

SIZE	⅛	¼	⅜	½	¾	1	1¼	1½	2	2½	3	3½	4 in.
Malleable Iron	....	.15	.18	.20	.28	.34	.46	.60	.80	1.50	2.10	3.00	4.00
" Galvanized	....	.20	.24	.27	.37	.50	.70	.90	1.20	2.25	2.90	4.50	5.60
Brass, Ground	.35	.40	.55	.75	1.00	1.40	1.90	2.75	4.00	6.00	8.50	......	......

American or Keystone Unions.

SIZE	¼	⅜	½	¾	1	1¼	1½	2	2½	3 in.
Plain	.20	.24	.28	.35	.40	.56	.80	.95	2.00	2.75
Galvanized	.24	.28	.35	.46	.55	.78	1.12	1.35	2.90	3.75

Flange Unions.

SIZE	½	¾	1	1¼	1½	2	2½	3	3½	4	4½	5	6	7	8	10 in.
Iron	.60	.65	.70	.85	1.15	1.50	1.75	2.25	2.75	3.15	4.50	5.00	6.50	8.00	10.00	15.00

Couplings.

SIZE	⅛	¼	⅜	½	¾	1	1¼	1½	2	2½	3	3½	4	4½	5	6	7	8	10 in
Plain Wrought Iron	.05	.05	.06	.07	.10	.13	.17	.21	.28	.40	.60	.80	1.00	1.50	1.65	2.40	3.25	4.25	7.50
Galvanized "	....	....	.08	.10	.13	.18	.25	.32	.40	.55	.80	1.05	1.40	2.00	2.25	3.25	....	....	....
Right and Left Couplings	....	.07	.09	.12	.15	.19	.25	.32	.45	.65	1.00	....	....	....	....	....	....	....	....
Galvanized R. & L. or Reducing	....	.06	.08	.10	.17	.25	.35	.55	.75	1.30	2.00	3.00	4.00	5.50	6.00	8.00	16.00	20.00	30.00
Brass, Rough	.08	.10	.15	.20	.30	.40	.60	.75	1.25	2.00	3.00	....	....	....	....	....	....	....	....
Brass Reducing Couplings	....	....	.20	.25	.35	.40	.60	.75	1.55	....	....	....	....	....	....	....	....	....	....
Finished Brass "	.13	.15	.20	.25	.35	.45	.70	.85	1.55	2.50	3.00	....	....	....	....	....	....	....	....

Bushings.

SIZE	¼	⅜	½	¾	1	1¼	1½	2	2½	3	3½	4	4½	5	6	7	8	10in
Iron	.04	.05	.06	.07	.09	.13	.17	.27	.42	.60	.80	1.00	1.50	1.85	2.50	3.75	5.50	7.50
" Galvanized	...	.06	.07	.10	.14	.21	.30	.44	.59	.90	....	....	....	....	....	....	....	....
Brass	07	.09	.13	21	.38	.50	.67	.84	1.50	2.50	....	....	....	....	....	....	....	....

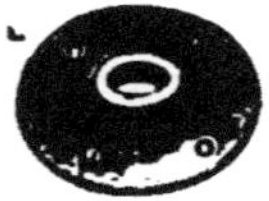

Flanges.

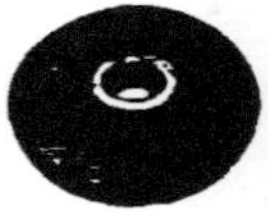

SIZE OF PIPE...	3/8	1/2	3/4	1	1 1/4	1 1/2	2	2 1/2	3	3 1/2	4	4 1/2	5	6	7	8	10
Diameter, 3 inches,	.14	.....	.....	.....	.....	.....	.....	.....	.....	.....	.....	.....	.....	.....	.....	.....	.....
" 4 "	.20	.20	.21	.22	.25	.....	.....	.....	.....	.....	.....	.....	.....	.....	.....	.....	.....
" 4 1/2 "	.26	.26	.28	.28	.30	.31	.....	.....	.....	.....	.....	.....	.....	.....	.....	.....	.....
" 5 "	.31	.31	.33	.33	.35	.36	.38	.....	.....	.....	.....	.....	.....	.....	.....	.....	.....
" 5 1/2 "	.40	.40	.40	.42	.42	.42	.45	.45	.....	.....	.....	.....	.....	.....	.....	.....	.....
" 6 "	.50	.50	.50	.52	.52	.52	.55	.55	.55	.....	.....	.....	.....	.....	.....	.....	.....
" 6 1/2 "	.....	.55	.60	.62	.62	.62	.65	.65	.65	.70	.....	.....	.....	.....	.....	.....	.....
" 7 "	.....	.65	.68	.72	.72	.72	.75	.75	.75	.75	.....	.....	.....	.....	.....	.....	.....
" 7 1/2 "	.....	.75	.80	.80	.80	.80	.80	.80	.84	.87	.87	.....	.....	.....	.....	.....	.....
" 8 "	.....	.85	.90	.90	.90	.90	.90	.90	.96	1.00	1.04	.....	.....	.....	.....	.....	.....
" 8 1/2 "	.....	.95	1.00	1.00	1.00	1.00	1.00	1.00	1.08	1.13	1.22	.....	.....	.....	.....	.....	.....
" 9 "	.....	.....	.....	1.10	1.10	1.10	1.15	1.15	1.22	1.26	1.40	1.55	1.65	.....	.....	.....	.....
" 9 1/2 "	.....	.....	.....	.....	1.20	1.20	1.25	1.30	1.37	1.55	1.58	1.70	1.80	2.20	.....	.....	.....
" 10 "	.....	.....	.....	.....	1.40	1.40	1.45	1.45	1.52	1.75	1.76	1.90	2.00	2.40	.....	.....	.....
" 11 "	.....	.....	.....	.....	.....	.....	1.90	1.90	1.90	1.95	2.16	2.32	2.40	2.80	2.80	.....	.....
" 12 "	.....	.....	.....	.....	.....	.....	.....	2.25	2.35	2.50	2.56	2.76	2.80	3.20	3.75	4.00	.....
" 13 "	.....	.....	.....	.....	.....	.....	.....	.....	2.60	2.85	2.85	3.00	3.05	3.45	4.10	4.50	.....
" 14 "	.....	.....	.....	.....	.....	.....	.....	.....	3.25	3.25	3.50	3.75	3.75	3.75	4.50	5.00	6.00
" 15 "	.....	.....	.....	.....	.....	.....	.....	.....	.....	.....	.....	.....	.....	.....	5.00	5.60	6.60
" 16 "	.....	.....	.....	.....	.....	.....	.....	.....	.....	.....	.....	.....	.....	.....	5.50	6.25	7.25

Wrought Iron Nipples.

LENGTH, INCHES.						SIZE.	PRICES.		PRICES EXTRA LONG.							
									Lengths of Extra Long, Inches							
Close.	Short	Long.				Inch.	Close or Short	Long.	5	6	7	8	9	10	11	12
3/4	1½	2	2½	3	3½	1/8	.05	.07	.16	.17	.18	.19	.20	.21	.22	.23
7/8	1½	2	2½	3	3½	1/4	.05	.07	.16	.17	.18	.19	.20	.21	.22	.23
1	1½	2	2½	3	3½	3/8	.06	.09	.17	.18	.19	.20	.21	.22	.23	.25
1 1/8	1½	2	2½	3	3½	1/2	.07	.10	.18	.19	.20	.21	.22	.23	.25	.27
1 3/8	2	2½	3	3½	4	3/4	.09	.11	.20	.21	.22	23	.25	.27	.29	31
1 1/2	2	2½	3	3½	4	1	.10	.15	.22	.24	.27	.29	.31	.33	.36	.40
1 5/8	2½	3	3½	4	4½	1¼	.14	.20	.29	.31	.33	.35	.38	40	.43	.46
1 3/4	2½	3	3½	4	4½	1½	.17	.25	.36	.38	.40	.42	.45	.48	.51	.55
2	2½	3	3½	4	4½	2	.25	.35	.44	.49	.54	.59	.64	.69	.74	.79
2 1/2	3	3½	4	4½	5	2½	.56	.75	……	.80	.85	.91	1.00	1.10	1.20	1.30
2 1/2	3	3½	4	4½	5	3	.75	.95	……	1.00	1.06	1.15	1.24	1.34	1.44	1.55
2 3/4	4	4½	5	5½	6	3½	1 00	1.25	……	……	1.38	1.50	1.62	1.74	1.86	2.00
3	4	4½	5	5½	6	4	1.25	1.60	……	……	1.75	1.92	2.10	2.30	2.50	2.70
3	4	4½	5	5½	6	4½	1.75	2.25	……	……	2.35	2.45	2.56	2.75	2.94	3.15
3 1/2	4½	5	5½	6	6½	5	2.00	2.60	……	……	2.75	2.95	3.20	3.45	3.80	4.20
3 1/2	4½	5	5½	6	6½	6	2.75	3.60	……	……	3.75	3.90	4.15	4.40	4.65	4.90
4	5	……	……	……	……	7	4.00	……	……	4.20	4.45	4.70	5.10	5.50	5.90	6.30
4	5	……	……	……	……	8	5.75	……	……	6 00	6.30	6.60	6.90	7.20	7.50	7.80
……	……	……	……	……	……	9	7.00	……	……	……	7.40	7.80	8.20	8.80	9.40	10 00
……	……	……	……	……	……	10	8.50	……	……	……	9.00	9.60	10.20	11.00	11.80	12.60
……	……	……	……	……	……	12	12.00	……	……	……	13.00	14.25	15.50	16.75	18.00	19.25

Wrought Iron Nipples, Right and Left.

SIZE……	1/4	3/8	1/2	3/4	1	1¼	1½	2	2½	3	3½	4 in.
Short……	.10	.10	.12	.15	.18	.24	.30	.40	1.00	1.25	1.50	1.75
Long……	.12	.14	.16	.20	.24	.35	.46	60	1.30	1.60	2.00	2.40

Wrought Iron Nipples, Galvanized.

SIZE	¼	⅜	½	¾	1	1¼	1½	2	2½	3	3½	4 in.
Short	.07	.08	.09	.11	.13	.17	.23	.33	.65	1.00	1.25	1.45
Long	.09	.11	.13	.16	.19	.24	.31	.40	.85	1.20	1.50	1.90

Brass Nipples.

SIZE	⅛	¼	⅜	½	¾	1	1¼	1½	2	2½	3 in.
Short	.12	.15	.20	.25	.30	.40	.60	.90	1.25	2.50	5.00
Shoulder	.15	.20	.30	.35	.45	.60	.90	1.25	1.60	3.00	6.00
Finished Brass	.15	.20	.30	.35	.45	.60	.90	1.25	1.60	3.00	6.00

For price of Copper Nipples add 25 per cent to above prices.

Standard Threads for Wrought Iron Pipe.

SIZE	⅛	¼	⅜	½	¾	1	1¼	1½	2	2½	3	3½	4	4½	5	6 in.
	.05	.05	.05	.05	.05	.06	.07	.08	.10	.15	.20	.25	.35	.45	.55	.70

Price for Cutting only, same as above. No extra charge will be made for Cutting when threads are made.

Plugs.

SIZE	⅛	¼	⅜	½	¾	1	1¼	1½	2	2½	3	3½	4	4½	5	6	7	8	10in
Iron	....	.03	.03	.04	.05	.06	.10	.13	.20	.35	.50	.75	.85	1.35	1.75	2.40	3.75	5.50	7.50
Galvanized	....	.05	.05	.06	.08	.10	.15	.23	.35	.57	.95	1.35	1.60	2.35	3.45	4.65	..	..	...
Brass, Rough	.06	.08	.10	.15	.20	.25	.40	.50	.90	1.50	2.25	3.00	4.00	....	...	...	....	....	...
" Finished	.10	.12	.15	.20	.25	.30	.50	.70	1.20	2.00	3.00	....	...	....	...	...	....	....	...

Caps.

SIZE	⅛	¼	⅜	½	¾	1	1¼	1½	2	2½	3	3½	4	4½	5	6	7	8	10 in.
Iron	.03	.03	.04	.05	.08	.12	.16	.24	.32	.45	.85	1.00	1.20	1.60	2.00	2.35	4.00	4.35	7.25
Galvanized	...	....	.05	.08	.12	.17	.24	.38	.52	.76	1.30	....	...	....	..	....	..	...	......
Brass, Rough	.08	.10	.15	.20	.30	.40	.60	.75	1.25	2.00	3.00	.	..	....	..	..	..	...	......
Brass, Finished	.13	.15	.20	.25	.80	.45	.70	.85	1.55	2.50	3.75	...	..			..	..	..	......

Locknuts.

SIZE	¼	⅜	½	¾	1	1¼	1½	2	2½	3	3½	4	4½	5	6	7	8	10 in.
Iron	.02	.03	.04	.05	.07	.00	.11	.18	.40	.50	.70	.95	1.25	1.35	1.90	2.50	3.50	4.50
Galvanized	.03	.04	.05	.07	.10	.14	.20	.30	.60	.75	1.10	1.50	2.00	2.20	3.00	..	....	......
Brass, Rough	.08	.10	.15	.20	.25	.40	.56	.90	1.75	2.75	..	....	..	..	.	....	..	......
Brass, Finished	.10	.12	.18	.23	.30	.45	.55	1.00	1.90	3.00	..	....	..		..	....	..	....

Y Branches.

SIZE	....	....	....	½	¾	1	1¼	1½	2	2½	3	3½	4	5	6 in.
Iron	..	..	...	.25	.30	.40	.60	.90	1.25	2.25	3.25	4.50	6.00	9.00	12.00
Reducing	..	..	..	.29	.35	.46	.70	1.35	1.90	2.60	3.75	5.20	6.90	10.35	13.80

Long Screw Connections.

SIZE	.....	¼	⅜	½	¾	1	1¼	1½	2	2½	3	4 in.
Iron	.....	.30	.35	.40	.55	.75	1.00	1.30	1.70	2.70	3.70	6.60

American Long Screws.

Require no Packing and make a Tight Joint.

SIZE	½	¾	1	1¼	1½	2	2½	3
Length	3½	4	4½	5	5½	6	7	8
Plain	.45	.55	.75	1.00	1.50	2.00	3.37	4 50
Galvanized	.60	.75	1.00	1.35	2.00	2.70	4.50	6.00

Ceiling Plates.

SIZE...	½	¾	1	1¼	1½	2	2½	3	3½	4 in.
....	..	.14	.18	.24	.32	.40	.50	.75	..	....

Floor Plates.

SIZE....	....	½	¾	1	1¼	1½	2	2½	3	3½	4	....	5	6 in.
....	..	.08	.10	.12	.15	.20	.25	.35	.40	.50	.65	...	.80	1.00

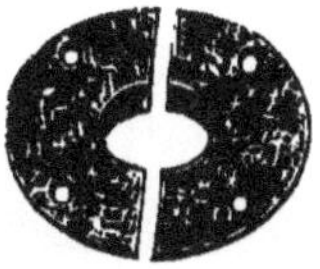

Floor Plates in Halves.

SIZE	½	¾	1	1¼	1½	2	2½	3	3½	4	4½	5	6
Price	.06	.07	.09	.12	.16	.20	.25	.35	.50	.65	.75	.80	1.00

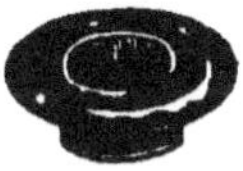

N. P. Floor and Ceiling Plates.

SIZE	¼	⅜	½	¾	1	1¼	1½	2
Price	.08	.08	.10	.10	.11	.12	.13	.15

Right and Left and Left Hand Fittings not mentioned will be charged 15 per cent. more than Right Hand Fittings.

Wrought Iron Pipe Hooks.

SIZE......	¼	⅜	½	¾	1	1¼	1½	2	2½	3	3½	4 in.
Price per 100. .	$1.00	1.00	1.00	1.50	1 75	2.00	2 50	3.00	5.00	6.00	10.00	12.50

Hook Plates.

No. OF HOOKS......	1	2	3	4	5	6	7	8	9	10	11	12
For ¾ inch Pipe......	.07	.12	.16	.20	.24	.28	.32	.36	.43	.50	.60	.70
" 1 " "	.09	.15	.21	.27	.32	.40	.48	.56	.65	.70	.80	1.00
" 1¼ " "	.10	.20	.30	.40	.50	.65	.75	.80	.95	1.05	1.15	1.25
" 1½ " "	.20	.40	.60	.80	1.00	1.20	1.40	1.60	1.80	2.00	2.20	2.40
" 2 " "	.30	.50	.75	1.00	1.25	1.50	....	..	...	....	...	....

Expansion Plates.

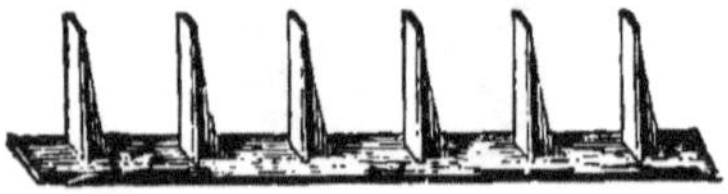

No. OF HOOKS..	2	3	4	5	6	7	8	10	12
For ¾ inch Pipe.	.16	.24	.34	.40	.50	.60	.70	.90	1.10
" 1 " "	.20	.27	.38	.45	.55	.65	.75	.95	1.20
" 1¼ " "	.30	.38	.50	.65	.75	.90	1.05	1.35	1.75

Ring Plates.

No. OF RINGS.......	2	3	4	5	6	8	10	12
For ¾ inch Pipe... ...	.22	.30	.40	.50	.60	.80	1.15	1.50
" 1 " "	.25	.35	.45	.55	.65	.85	1.20	1.60
" 1¼ " "	.40	.60	.90	1.00	1.15	1.40	......	.....

Combined Steel Pipe Hooks, Expansion Plates and Ring Plates.

NUMBER OF PIPES...		1	2	3	4	5	6	7	8	9	10	11	12
Hook Plates and Expansion Plates	¾ in	07	.12	.16	.20	.24	.28	.32	.36	.43	.50	.60	.70
	1 "	.09	.15	.21	.27	.32	.40	.48	.56	.65	.70	.80	1.00
	1¼ "	.10	.20	.30	.40	.50	.65	.75	.80	.95	1.05	1.15	1.25
	1½ "	.20	.40	.60	.80	1.00	1.20	1.40	1.60	1.80	2.00	2.20	2.40
	2 "	.30	.50	.75	1.00	1.25	1.50	....	....	....	....	....	....
Ring Plates and Pipe Hook	¾ "	.13	.22	.30	.40	.50	.60	.70	.80	.95	1.15	1.35	1.50
	1 "	.15	.25	.35	.45	.55	.65	.75	.85	1.00	1.20	1.40	1.60

Pipe Hangers.

SIZE......	⅜	½	¾	1	1¼	1½	2	2½	3	3½	4	5	6
....	.15	.15	.18	.18	.20	.22	.25	.30	.35	.37	.40	.45	.50

Roller Pipe Hangers.

No. OF PIPES	1	2	3	4	5	6
For 1 inch Pipe.........	.06	.12	.18	.24	.30	.36
" 1¼ " "	.07	.14	.21	.28	.35	.42
" 1½ " "	.08	.16	.24	.32	.40	.48
" 2 " "	.13	.26	.39	.52	.65	.78

Above Prices for Rolls only, without Rods, Bolts or Flanges.

Coil Stands.

No. OF PIPES HIGH........	4	6	8	10	12
¾ inch Pipe, per Pair..........	.55	.70	.85	1.25	1.50
1 " " "	.60	.75	1.30	1.60	2.05

Branch Tees or Manifolds.

No. of Outlets..... ..		2	3	4	5	6	7	8	9	10	12
¾ in. Outlets, 2 in. Centre, to Centre.	1 in. or 1¼ in. Run,	.50	.65	.80	.95	1.10	1.35	1.50	1.65	1.85	2.50
	1½ in. "	.65	.85	1.05	1.15	1.30	1.65	2.00	2.50	3.00	...
1 in. Outlets, 2½ in. Centre to Centre,	1 in. or 1¼ in. Run.	.70	.80	.95	1.10	1.35	2.05	2.35	2.55	2.85	3.75
	1½ "	.75	.90	1.05	1.20	1.50	2.20	2.50	2.80	3.15	4.00
	2 "	1.00	1.20	1.60	1.80	2.00	2.40	2.80	3.30	4.00	4.75
	2½ "	2.10	2.50	2.90	3.25	3.60	4.00	4.50	5.00	5.50	6.00
1¼ in. Outlets 3 in. Centre to Centre,	1½ in. Run......	1.20	1.60	2.00	2.40	2.80	3.20	3.60	4.00	4.40	4.80
	2 in. Run.......	1.40	1.85	2.45	2.99	3.40	3.90	4.40	5.00	5 50	6.00
1½ in. Outlets, 3½ in. Centre to Centre,	2 in. Run.......	1.75	2.25	2.75	3.25	3.75	4.25	4.75	5.50	6.00	6.50

Back or Side Outlets charged as Additional Front Outlets. Left Hand Tees Furnished only to Order.

Wrenches for Brass Cocks.

Size.......	¼	⅜	½	¾	1	1¼	1½	2	2½	3 in.
(Iron)......................	.10	.10	.12	.15	.02	.25	.30	.40	.75	1.00

Brass Finished Rail Fittings.

Size	1	1¼	1½	2
Fig. 700..................	.80	1.20	1.60	2.50
" 701......	1.10	1.70	2.00	3.00
" 702	1.10	1.70	2.00	3.00
" 703...	1.50	2.00	2.40	3.50
" 704	1.50	2.00	2.40	3.50
" 705.......	1.70	2.25	3.00	4.00
" 706..........................	.40	.70	1.00	1.30
" 707	2.00	2.50	3.25	4.50

For illustrations see page 19.

Malleable Iron and Brass Rail Fittings,

FOR FENCES, ENCLOSING ENGINES AND MACHINERY, EXHIBITION SPACES, &c.

Fig. 700.

Fig. 701.

Fig. 702.

Fig. 703.

Fig. 704.

Fig 705.

Fig. 706.

Fig. 707.

Please Read these Notes with care.

IN ordering these Railing Fittings give NUMBER of our figure, and state whether *right-hand or left-hand* threads are wanted. Where Fittings are required having BOTH RIGHT AND LEFT-HAND outlets, please fully describe which outlets are wanted RIGHT-HAND and which LEFT-HAND. A careful observance of the above will save much trouble, and secure the accurate filling of your orders.

To construct a Railing *two* pipes high, the upper outlets of all fittings used in lower pipe should be tapped with LEFT-HAND THREAD, but when orders are sent us without specifying how outlets are to be tapped, ALL FITTINGS WILL INVARIABLE BE FURNISHED RIGHT-HAND.

☞ As these Fittings do not need to be Steam or Water-tight, a sufficiently clean thread to screw up well and make a good job, can be made by running a LEFT-HAND TAP into any outlet tapped RIGHT-HAND.

Malleable Iron Rail Fittings.

PIPE SIZE		3/4	1	1 1/4	1 1/2	2
Fig. 700—Elbow	each	.15	.20	.35	.45	.72
" 701—Elbow, side outlet	"	.18	.25	.40	.50	.80
" 702—Tee	"	.18	.25	.40	.50	.75
" 703—Tee, side outlet	"	.22	.35	.45	.55	.90
" 704—Cross	"	.22	.35	.45	.58	1.00
" 705—Cross, side outlet	"	.27	.40	.50	.65	1.25
" 706—Floor Flange	"	.12	.15	.20	.28	.30
" 707—Acorn Ornament	"	.15	.20	.25	.35	.90

We also furnish brackets for counter-rails.

Special Cast Iron Fittings for Water Connections.

No. 1.

No. 2.

No. 3.

No. 4.

SIZE	1	1¼	1½	2	2½	3	3½	4	4½	5	6	7	8	10	12
No. 1, Price each	.25	.35	.45	.60	1.00	1.50	2.00	2.50	3.50	4.50	6.50	10.00	14.00	20.00	30.00
No. 2, " "	.38	.52	.68	.90	1.50	2.25	3.00	3.75	5.25	6.75	9.75	15.00	21.00	30.00	45.00
No. 3, " "	.38	.52	.68	.90	1.50	2.25	3.00	3.75	5.25	6.75	9.75	15.00	21.00	30.00	45.00
No. 4, " "	....	....	.90	1.20	2.00	3.00	4.00	5.00	.....	9.00	13.00	20.00	28.00	40.00	60.00

Medium Cast Iron Fittings for Water Connections.

No. 5.

No. 6.

No. 7.

No. 8.

SIZE	1	1¼	1½	2	2½	3	3½	4	5	6	7	8	10	12
No. 5, Price each	.25	.35	.45	.60	1.00	1.50	2.00	2.50	4.50	6.50	10.00	14.00	20.00	30.00
No. 6, " "	.33	.52	.68	.90	1.50	2.25	3.00	3.75	6.75	9.75	15.00	21.00	30.00	45.00
No. 7, " "	......	.80	1.00	1.35	2.25	3.50	.....	5.75	......	10.00	15.00	.....		
No. 8, " "	......	.80	1.00	1.35	2.25	3.50	4.50	5.75	......	10.00	15.00	31.00	......	......

Nos. 7 and 8 made in reducing only.

List of Standard Sizes of Cast Iron Fittings.

Sizes differing from standard sizes will be charged at 10 per cent. gross discount higher than standard sizes.

CAST IRON FITTINGS.

Flange Unions.

Size ..	½	¾	1	1¼	1½	2	2½	3	4	4½	5	6	7	8	9	10	12

Cast Iron Reducing Couplings.

Size.	Size.	Size.	Size.	Size.
2½ x 2	4 x 3½	4 x 2	5 x 3	8 x 6
3 x 2½	4 x 3	4½ x 4	6 x 5	10 x 8
3 x 2	4 x 2½	5 x 4	6 x 4	12 x10
3½ x 3				

Cast Iron Caps.

Size......	2½	3	3½	4	4½	5	6	7	8	10	12

Elbows.

Size.	Size.	Size.	Size.	Size.	Size.	Size.	Size.
¼ x ¼	¾ x ½	1¼ x 1	1½ x 1	2 x 1	3 x 2½	4 x 3½	7 x 7
⅜ x ⅜	1 x 1	1¼ x ¾	1½ x ¾	2½ x 2½	3 x 2	4 x 3	8 x 8
½ x ½	1 x ¾	1¼ x ½	2 x 2	2½ x 2	3½ x 3½	4½ x 4½	9 x 9
¾ x ¾	1 x ½	1½ x 1½	2 x 1½	2½ x 1½	3½ x 3	5 x 5	10 x10
½ x ⅜	1¼ x 1¼	1½ x 1¼	2 x 1¼	3 x 3	4 x 4	6 x 6	12 x12

45° Elbows.

Size.	⅜	½	¾	1	1¼	1½	2	2½	3	3½	4	4½	5	6	7	8	9	10	12

Bushings.

Reducing, one size only, are malleable up to 2½ in.

Size.	Size.	Size.	Size	Size.	Size.	Size.	Size.
⅜ x ¼	1¼ x 1	2 x 1	3 x 2	4 x 3½	4½ x 3	6 x 4½	8 x 7
½ x ⅜	1¼ x ¾	2 x ¾	3 x 1½	4 x 3	4½ x 2½	6 x 4	8 x 6
½ x ¼	1¼ x ½	2 x ½	3 x 1¼	4 x 2½	5 x 4½	6 x 3½	8 x 5
¾ x ½	1¼ x ⅜	2½ x 2	3 x 1	4 x 2	5 x 4	6 x 3	8 x 4
¾ x ⅜	1½ x 1¼	2½ x 1½	3½ x 3	4 x 1½	5 x 3½	6 x 2½	10 x 8
¾ x ¼	1½ x 1	2½ x 1¼	3½ x 2½	4 x 1¼	5 x 3	7 x 6	10 x 6
1 x ¾	1½ x ¾	2½ x 1	3½ x 2	4 x 1	5 x 2½	7 x 5	12 x10
1 x ½	1½ x ½	2½ x ¾	3½ x 1½	4½ x 4	5 x 2	7 x 4½	12 x 8
1 x ⅜	2 x 1½	3 x 2½	3½ x 1¼	4½ x 3½	6 x 5	7 x 4	
1 x ¼	2 x 1¼						

Tees.

In describing Tees the run is first named, then the outlet.

Size.	Size.	Size.	Size.	Size.
¼ x ¼ x ¼	1¼ x 1 x 1½	2 x 1½ x 1¼	3 x 1 x 3	4 x 4 x 5
⅜ x ⅜ x ⅜	1¼ x ¾ x 1½	2 x 1½ x 1	3 x 3 x 4	4½ x 4½ x 4½
½ x ½ x ½	1¼ x 1¼ x 2	2 x 1½ x ¾	3½ x 3½ x 3½	5 x 5 x 5
½ x ½ x ⅜	1½ x 1½ x 1½	2 x 1¼ x 2	3½ x 3½ x 3	5 x 5 x 4
½ x ½ x ¼	1½ x 1½ x 1¼	2 x 1¼ x 1½	3½ x 3½ x 2½	5 x 5 x 3½
¾ x ¾ x ¾	1½ x 1½ x 1	2½ x 2½ x 2½	3½ x 3½ x 2	5 x 5 x 3
¾ x ¾ x ⅜	1½ x 1½ x ¾	2½ x 2½ x 2	3½ x 3½ x 1½	5 x 5 x 2½
¾ x ¾ x ½	1½ x 1½ x ½	2½ x 2½ x 1½	3½ x 3½ x 1¼	5 x 5 x 2
¾ x ½ x ¾	1½ x 1¼ x 1½	2½ x 2½ x 1¼	3½ x 3 x 3	5 x 4 x 5
¾ x ½ x ½	1½ x 1¼ x 1¼	2½ x 2½ x 1	3½ x 3 x 2½	5 x 4 x 4
¾ x ⅜ x ¾	1½ x 1¼ x 1	2½ x 2½ x ¾	3½ x 3 x 2	5 x 3 x 5
¾ x ¾ x 1	1½ x 1¼ x ¾	2½ x 2 x 2½	3½ x 2½ x 3	5 x 2½ x 5
¾ x ½ x 1	1½ x 1¼ x ½	2½ x 2 x 2	3½ x 2½ x 2½	5 x 5 x 6
¾ x ¾ x 1¼	1½ x 1 x 1½	2½ x 2 x 1½	3½ x 2½ x 2	6 x 6 x 6
¾ x ¾ x 1½	1½ x 1 x 1¼	2½ x 2 x 1¼	4 x 4 x 4	6 x 6 x 5
¾ x ¾ x 2	1½ x 1 x 1	2½ x 2 x 1	4 x 4 x 3½	6 x 6 x 4
1 x 1 x 1	1½ x 1 x ¾	2½ x 1½ x 2½	4 x 4 x 3	6 x 6 x 3½
1 x 1 x ¾	1½ x 1 x ½	2½ x 1½ x 2	4 x 4 x 2½	6 x 6 x 3
1 x 1 x ½	1½ x ¾ x 1¼	2½ x 1½ x 1½	4 x 4 x 2	6 x 6 x 2½
1 x 1 x ⅜	1½ x ¾ x 1½	2½ x 1¼ x 2½	4 x 4 x 1½	6 x 6 x 2
1 x ¾ x 1	1½ x ½ x 1¼	2½ x 1¼ x 2	4 x 4 x 1¼	6 x 5 x 6
1 x ¾ x ¾	1½ x 1½ x 2	2½ x 1 x 2½	4 x 4 x 1	6 x 5 x 5
1 x ¾ x ½	1½ x 1¼ x 2	2½ x 2½ x 3	4 x 4 x ¾	6 x 4 x 6
1 x ½ x 1	1½ x 1 x 2	2½ x 2 x 3	4 x 3½ x 3½	6 x 3 x 6
1 x ½ x ¾	1½ x ¾ x 2	3 x 3 x 3	4 x 3½ x 3	7 x 7 x 7
1 x ½ x ½	2 x 1¼ x 1¼	3 x 3 x 2½	4 x 3½ x 2½	7 x 7 x 6
1 x ⅜ x 1	2 x 1¼ x 1	3 x 3 x 2	4 x 3 x 4	7 x 7 x 5
1 x 1 x 1¼	2 x 1¼ x ¾	3 x 3 x 1½	4 x 3 x 3½	7 x 7 x 4
1 x ¾ x 1¼	2 x 1 x 2	3 x 3 x 1¼	4 x 3 x 3	8 x 8 x 8
1 x 1 x 1½	2 x 1 x 1½	3 x 3 x 1	4 x 3 x 2½	8 x 8 x 6
1 x 1 x 2	2 x 1 x 1¼	3 x 3 x ¾	4 x 3 x 2	8 x 8 x 5
1 x ¾ x 2	2 x 1 x 1	3 x 2½ x 3	4 x 3 x 1½	8 x 8 x 4
1¼ x 1¼ x 1¼	2 x ¾ x 2	3 x 2½ x 2½	4 x 3 x 1¼	8 x 8 x 3
1¼ x 1¼ x 1	2 x ¾ x 1½	3 x 2½ x 2	4 x 2½ x 4	8 x 8 x 2½
1¼ x 1¼ x ¾	2 x ½ x 2	3 x 2½ x 1½	4 x 2½ x 3	8 x 8 x 2
1¼ x 1¼ x ½	2 x 2 x 2½	3 x 2½ x 1¼	4 x 2½ x 2½	9 x 9 x 9
1¼ x 1 x 1¼	2 x 2 x 3	3 x 2½ x 1	4 x 2½ x 2	10 x10 x10
1¼ x 1 x 1	2 x 2 x 2	3 x 2 x 3	4 x 2½ x 1	10 x10 x 8
1¼ x 1 x ¾	2 x 2 x 1½	3 x 2 x 2½	4 x 2 x 4	10 x10 x 6
1¼ x 1 x ½	2 x 2 x 1¼	3 x 2 x 2	4 x 2 x 3	10 x10 x 4
1¼ x ¾ x 1¼	2 x 2 x 1	3 x 2 x 1½	4 x 2 x 2½	12 x12 x12
1¼ x ¾ x 1	2 x 2 x ¾	3 x 1½ x 3	4 x 2 x 2	12 x12 x10
1¼ x ¾ x ¾	2 x 2 x ½	3 x 1½ x 2½	4 x 1½ x 4	12 x12 x 8
1¼ x ½ x 1¼	2 x 1½ x 2	3 x 1½ x 2	4 x 1¼ x 4	12 x12 x 6
1¼ x 1¼ x 1½	2 x 1½ x 1½	3 x 1¼ x 3	4 x 1 x 4	

Crosses.

Size.	Size.	Size.	Size.
½ x ½ x ½ x ½	1½ x 1½ x ¾ x ¾	2½ x 2½ x 1 x 1	4 x 4 x 3½ x 3½
¾ x ¾ x ¾ x ¾	2 x 2 x 2 x2	3 x 3 x 3 x 3	4 x 4 x 3 x 3
1 x 1 x 1 x 1	2 x 2 x 1½ x1½	3 x 3 x 2½ x 2½	4½ x 4½ x 4½ x 4½
1 x 1 x ¾ x ¾	2 x 2 x 1¼ x1¼	3 x 3 x 2 x 2	5 x 5 x 5 x 5
1¼ x 1¼ x 1¼ x 1¼	2 x 2 x 1 x1	3 x 3 x 1½ x 1½	6 x 6 x 6 x 6
1¼ x 1¼ x 1 x 1	2½ x 2½ x 2½ x 2½	3½ x 3½ x 3½ x 3½	7 x 7 x 7 x 7
1½ x 1½ x 1½ x 1½	2½ x 2½ x 2 x 2	3½ x 3½ x 3 x 3	8 x 8 x 8 x 8
1½ x 1½ x 1¼ x 1¼	2½ x 2½ x 1½ x 1½	3½ x 3½ x 2½ x 2½	10 x10 x10 x10
1½ x 1½ x 1 x 1	2½ x 2½ x 1¼ x 1¼	4 x 4 x 4 x 4	12 x12 x12 x12

Classification of Malleable Iron Fittings.

CLASS A. 25c. per lb.
Elbows, 1/8, 1/4 × 1/8, 3/8 × 1/4.
Tees, 1/8, 1/8 × 1/4, 1/4 × 1/8, 3/8 × 1/8.
Reducing Couplings, 1/4 × 1/8, 3/8 × 1/8.
R. H. Couplings, and R. & L. Couplings, 1/8.
Rod Couplings.

CLASS B. 16c. per lb.
Elbows and Tees. 1/4 to 1/2 inch. inclusive.
Elbows, Side Outlets, all sizes.
Street Elbows 3/4 and smaller.
Crosses, 1 inch and smaller.
Drop **L**'s and **T**'s.
Four Way Tees.
Caps and Locknuts, 1 inch and smaller.
Reducing Couplings, 3/8 × 1/4 to 1 inch, inclusive.
R. and L. Couplings, 1/4 to 3/4, inclusive.
Extension Pieces, all sizes.
R. and L. Fittings, 1 inch and smaller.
R. H. Couplings, 1/4 to 3/4, inclusive.
Waste Nuts.
Chandelier Hooks and Loops.
Return Bends, to 1 inch.
Wall Plates.

CLASS C. 13c. per lb.
Elbows and Tees, 3/4 to 1 inch; such as have smaller hole than 3/4 to be classed as class B.
St. Elbows, 1 inch and larger.
Crosses, all sizes, 1 1/4 and larger.
Caps and Locknuts, all sizes, 1 1/4 and larger.
Reducing Couplings, all sizes, 1 1/4 and larger.
R. & L. Fittings, all sizes, 1 1/4 and larger.
R. H. Couplings, 1 inch and 1 1/4.
Return Bends, 1 1/4 and larger.
R. and L. Couplings, 1 inch and larger.

CLASS D. 11c. per lb.
Elbows and Tees, above 1 inch; such Fittings in this class that have holes *smaller* than 1 inch to be classed as class C.

Galvanized Fittings, Standard List.

CLASS...............	A.	B.	C.	D.
Price per lb..........	.35	.23	.20	.18c.

An extra charge of 10c. per lb. will be added to price of Galvanized Fittings not enumerated in Standard List.

NOTE.—All sizes, 2-inch and under, furnished either plain for Gas or beaded for Steam.

All sizes above 2-inch have bead or band.

In ordering be particular to mention whether for Gas or Steam. The cuts on next page show the different kinds.

Gas. Steam.

ELBOWS.

Gas. Steam.

TEES.

STREET ELBOWS.

MALE AND FEMALE SCREW.

Gas. Steam.

CROSSES.

ELBOWS WITH SIDE OUTLETS.

DROP ELBOWS.

FEMALE.

DROP ELBOWS.

MALE AND FEMALE.

DROP ELBOWS.

WITH LONG OUTLET PIECE.

DROP ELBOWS.

FLANGES RIGHT OR LEFT.

DROP TEES.

FEMALE.

LOCK-NUTS.

DROP TEES.

MALE AND FEMALE.

CAPS.

COUPLINGS.

RIGHT AND LEFT.

WASTE-NUTS.

TINNED.

REDUCING COUPLINGS.

COUPLINGS.

CHANDELIER HOOKS.

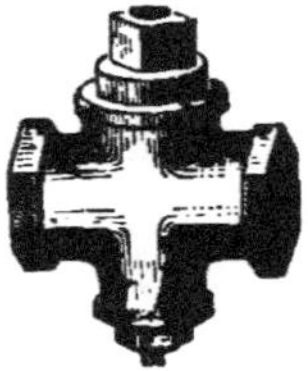

Iron Cocks.

SIZE	¼	⅜	½	¾	1	1¼	1½	2	2½	3	3½	4	4½	5	6 in.
All Iron	$.70	.75	.80	.90	1.25	1.50	2.00	2.60	4.50	6.50	12.00	16.00	....	33.00	45.00
" with Brass Washers	....	....	.90	1.00	1.40	1.75	2.25	3.00	5.00	7.50	14.00	19.50	....	39.00	53.00
" Flanged	....	....	....	1.65	2.25	2.75	3.50	4.35	6.50	9.50	15.50	20.00	....	37.00	50.00
" with Brass Plugs, Screwed	1.00	1.10	1.20	1.60	2.00	2.75	4.00	5.00	9.50	13.50	30.00	40.00	....	70.00	95.00
" " " Flanged	....	....	....	2.35	3.00	4.00	5.50	6.75	11.50	16.50	33.50	44.00	....	75.00	101.00
" Three Way, Screwed	....	....	....	1.30	1.75	2.00	2.75	4.00	6.00	8.50	15.00	20.00	....	40.00	55.00
" " Flanged	....	....	....	2.30	3.25	3.75	5.00	6.50	9.00	13.00	20.25	26.00	....	46.00	61.00
" " with Brass Plugs Screwed	....	....	....	2.00	2.50	3.25	4.75	6.50	11.00	15.50	33.00	44.00	....	70.00	95.00
" " " " Flanged	....	....	....	3.00	4.00	5.00	7.00	9.00	14.00	20.00	38.25	50.00	....	80.00	105.00

Standard Dimensions of Iron Body Flanged Valves.

SIZE	2	2½	3	3½	4	4½	5	6	7	8	10	12 in
Globe Valves, Diameter of Flanges	6½	7	8	9	10	10	11	12	13	14	16	19
" " Distance, face to face	5⅝	7⅜	7½	9⅜	11	11	12	13¾	16	16½	19¾	22½
Angle " Diameter of Flanges	6½	7	8	9	10	10	11	12	13	14	16	19
" " Distance, centre to faces	3½	3¾	4⅝	5⅜	5⅞	5⅞	6⅜	7⅛	7¾	8¼	9½	11¼
Safety " Diameter of Flanges	6½	7	8	9	10	10	11	12	13	14	16	19
" " Distance, face to face	7	8½	9½	9½	11	11½	12	13¾	....	16	....	....
" " " centre to inlet	3½	4¼	4⅝	5	5½	5⅞	6⅜	7⅛	....	8	....	....
Jenkins' Globe Valves, Diameter of Flanges	6¼	7	8	9	10	10	11	12	13	14	16	19
" " " Distance, face to face	6¼	7¼	9	10	11½	12½	13	16	16	18½	21¼	24¼
" Angle " Diameter of Flanges	6½	7	8	9	10	10	11	12	13	14	16	19
" " " Distance, centre to face	3⅞	4	4⅝	5	5⅞	5⅞	6½	8	8½	9¼	11	12¼
" Gate " Diameter of Flanges	6½	7	7	8½	9	....	10	11	12	13	16	18
" " " Distance, face to face	6	7¼	7⅝	7⅝	8⅜	....	9½	10	11	12	14¼	14½

IRON BRASS MOUNTED GOODS.

Globe and Angle Valves.

SIZE	1	1¼	1½	2	2½	3	3½	4	4½	5	6	7	8	10 in.
Screwed Ends	2.00	2.50	3.50	5.00	7.50	10.50	......	......	......	......	......	......	......	......
Flanged "	3.00	3.75	5.00	6.75	9.50	13.50	......	......	......	......	......	......	......	......
Screwed, with Yoke	......	......	......	8.00	10.50	14.50	18.00	21.00	28.00	32.00	44.00	75.00	85.00	135.00
Flanged " "	......	......	......	9.75	12.50	17.50	21.50	25.00	32.00	36 00	49.00	80.00	91.00	145 00

Horizontal Check Valves.

SIZE	1	1¼	1½	2	2½	3	3½	4	4½	5	6	7	8
Screwed ends	1.50	2.25	2.75	3.75	6.25	9.75	12.75	15.00	20.00	24.00	33.00	55.00	65.00
Flanged ends	2.50	3.50	4.25	5.50	8.25	12.75	16.25	19.00	24 00	28.00	38.00	60.00	71.00

Vertical Check Valves.

SIZE	2	2½	3	3½	4	4½	5	6	7	8 in.
Screwed ends	4.75	7.50	11.25	14.50	17.00	22.00	26.50	36.00	58.00	69.00
Flanged ends	6 50	9.50	14.25	18.00	21.00	26 00	30.50	41.00	63.00	75.00

Swing Check Valves.

Iron Body, Brass or Leather Discs, Brass Seat.

SIZE	2½	3	3½	4	5	6	7	8	10	12 in.
Price, Screwed Ends	10.00	12.00	16.00	18.00	25.00	32.00	41.00	50.00	65.00	95.00
" Flanged "	12.00	14.50	19.00	21.00	29.00	37.00	46.00	56.00	75.00	105.00
" Hub "	....	....	....	20.00	....	35.00	....	55.00	....	....
Diameter of Flanges	7	8	8½	9	10	11	13	14	16	19
Length, face to face, of flanges	6⅜	7½	8⅜	9⅜	10¾	12¼	14	16½	18¾	22
Distance from end to end, screwed	6¼	7¾	8½	9¾	11	12½	14	15¾	18¾	22

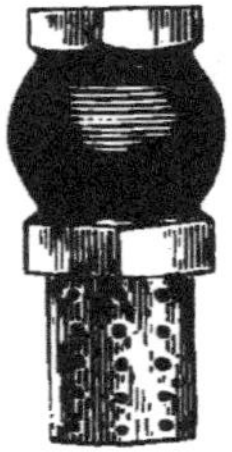

Foot Valves.—With Strainer.

SIZE	¾	1	1¼	1½	2	2½	3	3½	4	4½	5	6	8 in.
Screwed	1.25	1.50	1.75	2.50	3.25	4.25	5.50	7.50	10.00	12.00	13.00	24.00	50.00
Flanged	2.00	2.50	3.00	4.00	5.00	6.25	8.50	11.00	14.00	16.00	17.00	29.00	56.00

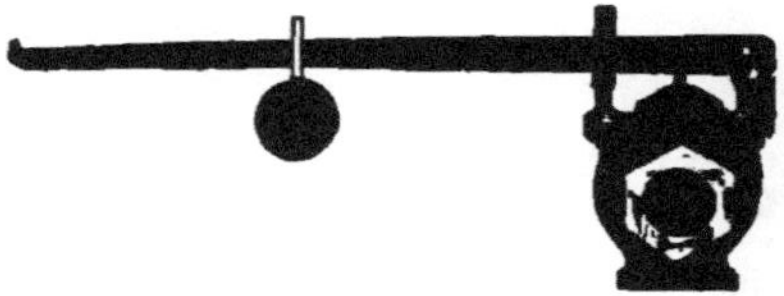

Globe and Angle Safety Valves.

SIZE	¾	1	1¼	1½	2	2½	3	3½	4	4½	5	6	8	10 in.
Screwed Ends	2.50	3.50	5.00	6.00	8.00	13.00	18.00	24.00	30.00	36.00	44.00	60.00	145.00	......
Flanged "	3.50	5.00	6.75	8.25	10.50	16.00	22.50	29.25	36.00	42.00	50.00	67.50	154.00	......

Back Pressure Valves.

SIZE	1½	2	2½	3	3½	4	4½	5	6	7	8	10 in.
Screwed Ends	7.00	8.00	10.50	14.50	18.00	21.00	28.00	32.00	44.00	75.00	85.00	135.00
Flanged Ends	8.50	9.75	12.50	17.50	21.50	25.00	32.00	36.00	49.00	80.00	91.00	145.00

Improved Noiseless Back Pressure Valves.

Flanged.

SIZE	4	6	8	10	12	14 in.
Price, each	25.00	49.00	91.00	145.00	220.00	350.00
Diameter of Flanges, Inches	9	11	14	16	19	21
Distance, face to face, of Flanges	9	12½	15½	19	22	26

Iron Body Butterfly Valves.

Brass Trimmings.

SIZE	2	2½	3	3½	4	5	6	8	10	12 in.
Price, Screwed, each	6.00	8.00	12.00	16.00	20.00	40.00	75.00	......	......	......
" Flanged, each	7.75	10.00	15.00	19.50	24.00	45.00	80.00	100.00	125.00	160.00

Iron Body Throttle Valves.

Brass Mounted.

SIZE	2½	3	3½	4	SIZE	2½	3	3½	4 in.
Price, Screwed, each	24.00	32.00	40.00	48.00	Price, Flanged, each	26 00	35 00	43.50	52.00

Standard Traverse Iron Body B. M. Expansion Joints.

Screwed.

SIZE	2	2½	3	3½	4	5	6	7	8	9	10	12 in.
Traverse, Inches	2½	2½	2¾	3	3¼	4	5	6	7	7	7	
Price, each	6.25	7.25	9.25	12.50	16 00	33.00	40.00	70.00	90.00	105.00	150.00	220.08

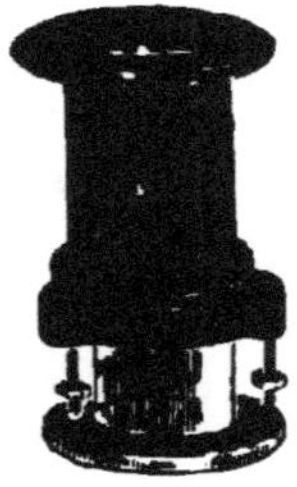

Standard Traverse Iron Body B. M. Expansion Joints.

Flanged.

Size	2	2½	3	3½	4	5	6	7	8	9	10	12 in.
Traverse	2½	2½	2¾	3	3¼	5	5	6	7	7	7	8
Diameter of Flanges	6½	7	8	8½	9	10	11	13	14	15	16	19
Price, each	13.25	14.25	16 75	22.50	26.00	44.00	50.00	82.00	105.00	120 00	170.00	240.00

When used for High Pressure Steam the expansion is 1 inch in 50 feet of pipe.

Price on 6, 10, 14, 14, 16 and 18 Traverses furnished on application.

STEAM BRASS WORK.

Globe and Angle Valves.

SIZE	1/8	1/4	3/8	1/2	3/4	1	1¼	1½	2	2½	3	3½	4 in.
Screwed	.60	.60	.75	1.00	1.35	1.80	2.80	3.90	5.90	11.25	16.00	30.00	40.00
Extra Heavy	.80	.80	.95	1.25	1.80	2.50	3.75	5.25	7.75	14.00	19.00	......	......
Flanged	......	......	......	3.50	4.00	5.00	7.00	9.00	14.00	20.00	30.00	44.00	65.00
Finished	1.75	1.75	2.00	2.25	2.75	3.50	5.00	7.00	10.00	......	......	......	......
Plated	2.10	2.10	2.40	2.65	3.20	4.00	5.60	7.65	10.90	......	......	......	......

Finished Globe Valves.

SIZE	1/8	1/4	3/8	1/2	3/4	1	1¼	1½	2 in.
Wood Wheel, Plain	1.30	1.30	1.60	1.85	2.15	2.85	4.00	5.50	8.50
Wood Wheel, Plated	1.55	1.55	1.90	2.15	2.50	3.25	4.45	6.00	9.25

Horizontal, Perpendicular and Angle Check Valves.

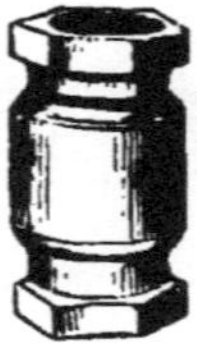

SIZE	1/8	1/4	3/8	1/2	3/4	1	1¼	1½	2	2½	3 in.
Screwed	.50	.50	.60	.85	1.15	1.55	2.30	3.25	5.20	10.00	14.00
Extra Heavy, Screwed	.65	.65	.75	1.15	1.60	2.00	3.25	4.50	6.75	12.50	17.00
Flanged	......	......	......	2.25	3.75	4.50	6.50	8.50	13.00	19.00	28.00

Union Check Valve.

Size	¼	⅜	½	¾	1	1¼	1½	2	2½	3 in.
Price, each, Finished	1.25	1.25	1.30	1.75	2.25	3.25	4.25	6.25	11.50	16.00

Cross Valves.

Size	¼	⅜	½	¾	1	1¼	1½	2	2½	3	3½	4	5 in.
Screwed	.85	1.00	1.50	2.00	2.50	3.50	5.00	8.00	16.00	24.00	....	....	....
Flanged	....	....	....	....	7.00	10.00	14.00	21.00	30.00	45.00	60.00	90.00	....

Globe and Angle Safety Valves.

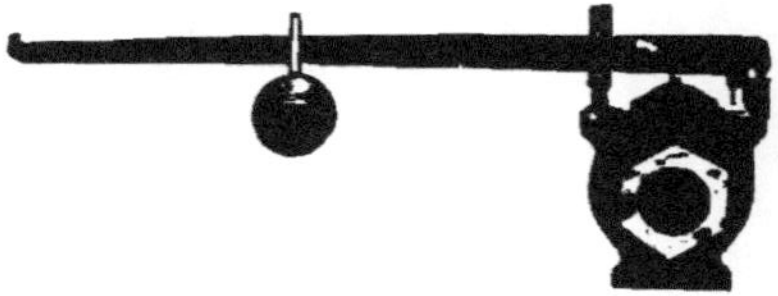

Size	¼	⅜	½	¾	1	1¼	1½	2	2½	3 in.
Screwed Ends	2.00	2.25	2.75	3.50	5.00	7.00	8.50	12.00	20.00	30.00
Flanged	....	....	....	....	9.50	13.50	17.50	25.00	34.00	50.00
Low Prssure Safety Valves	....	....	....	2.50	3.00	4.00	5.00	....	....	....

Hose Valves.

SIZE	1¼	1½	2	2½ in.
Male Threads	4.00	5.50	8.00	10.00

Brass Butterfly Valves.

SIZE	1	1¼	1½	2	2½	3 in.
Price, Each	3.50	4.50	5.50	8.00	11.00	16.00

Brass Throttle Valves.

SIZE	1	1¼	1½	2	2½ in.
Price each	9.00	11.00	13.00	20.00	30.00

Balance Valves.

SIZES	1	1¼	1½	2 in.
Price, each	3.50	5.00	7.00	10.00

Expansion Joints.

Standard Traverse.

SIZE.........Inches.	½	¾	1	1¼	1½	2	2½	3 in.
Traverse.........Inches.	2	2¼	2¼	2¼	2¼	2½	2½	2¾
Price	1.50	2.00	2.75	4.00	5.50	8.00	16.00	24.00

When used for high pressure steam, expansion is 1 inch in 50 ft. of pipe.
Price on 6, 8, 10 and 12 inch Traverse furnished on application.

Vacuum Valves.

SIZES.	¼	⅜	½	¾ in.
Each	.75	1.00	1.50	2.00

Radiator Valves.

With Wood Handles.

SIZE	½	¾	1	1¼	1½	2 in.
Rough Body, Plain	1.35	1.60	2.25	3.25	4.50	7.00
" " Nickel Plated	1.65	1.95	2.65	3.70	5.00	7.75
Finished " Plain	1.85	2.15	2.85	4.00	5.50	8.50
" " Nickel Plated	2.15	2.50	3.25	4.45	6.00	9.25
Rough " Plain, with Frink's Seat	1.65	1.95	2.65	3.85	5.35	8.00
" " Nickel Plated, with Frink's Seat	1.95	2.30	3.05	4.30	5.85	8.75
Finished " Plain " "	2.15	2.50	3.25	4.60	6.35	9.50
" " Nickel Plated " "	2.45	2.85	3.65	5.05	6.85	10 25

Radiator Valves.—Wood Wheel.

With Male or Female Unions.

SIZE	½	¾	1	1¼	1½	2 in.
Rough Body, Plain	2.05	2.45	3.25	4.50	6.50	10.00
" " Nickel Plated	2.40	2.85	3.65	5.05	7.10	10.85
Finished " Plain	2.55	3.00	3.85	5.25	7.50	11.50
" " Nickel Plated	2.90	3.40	4.80	5.80	8.10	12.35
Rough " Plain, with Frink Seat	2.35	2.80	3.65	5.10	7.35	11.00
" " Nickel Plated, with Firnk Seat	2.70	3.20	4.05	5.65	7.95	11.85
Finished " Plain, " "	2.85	3.35	4.25	5.85	8.35	12.50
" " Nickel Plated, " "	3.20	3.75	4.70	6.40	8.95	13.35

Corner Radiator Valves with Unions.

SIZE	¾	1	1¼	1½ in.
Rough Body, Brass Seat	3.50	4.30	5.85	7.75
" " Nickel Plated, Brass Seat	3.80	4.75	6.40	8.10
" " Frink Seat	3.85	4.75	6.45	8.55
Finished " " " "	4.40	5.30	7.05	9.65
Rough " Nickel Pl., Frink Seat	4.20	5.25	7.05	8.95
Finished " " " " "	4.70	5.80	7.70	10.20

Radiator Valves.

With Patent Cap Attachment.

SIZE	½	¾	1	1¼	1½ in.
Rough Body, Plain Brass Seat	1.80	2.10	2.90	4.00	5.50
Finished " " " "	2.30	2.65	3.50	4.75	6.50
Rough " Nickel Plated Brass Seat	2.10	2.45	3.30	4.45	6.00
Finished " " " " "	2.60	3.00	3.90	5.20	7.00
Rough " Plain Frink Seats	2.10	2.45	3.30	4.60	6.35
Finished " " " "	2.60	3.00	3.90	5.35	7.35
Rough " Nickel Plated "	2.40	2.80	3.70	5.05	6.85
Finished " " " "	2.90	3.35	4.30	5.80	7.85

Union Radiator Valves.

With Patent Cap Attachment.

SIZE	½	¾	1	1¼	1½ in.
Rough Body, Plain Brass Seat	2.50	2.95	3.90	5.25	7.50
Finished " " " "	3.00	3.50	4.50	6.00	8.50
Rough " Nickel Plated, Brass Seat	2.85	3.35	4.35	5.80	8.10
Finished " " " " "	3.35	3.90	4.95	6.55	9.10
Rough " Plain, Frink Seat	2.80	3.30	4.30	5.85	8.35
Finished " " " " "	3.30	3.85	4.90	6.60	9.35
Rough " Nickel Plated, Frink Seat	3 15	3.70	4.75	6.40	8.95
Finished " " " " "	3.65	4.25	5.35	7.15	9.95

Nickel Plated Rough Brass Union Radiator Elbows.

SIZE	¾	1	1¼	1½ in.
Price, each	2.00	2.50	3.20	4.00

Patent "Pedal Valve" for Radiators.

Number	Size, ...Inches.		3/4	1	1¼	1½
1	Rough Body, Finished Trimmings.	Square End cut right or left thread.	2.50	3.20	4.50	6.25
2	Finished all over..........		3.00	3.75	5.25	7.25
3	Rough Body, Plated Trimmings...		2.70	3.50	4.75	6.50
4	" " " all over.......		2.85	3.65	4.90	6.75
5	Finished and Plated all over		3.10	4.00	5.40	7.75
6	Rough Body, Finished Trimmings.	Ball Joint. Male Union.	3.50	4.30	5.85	7.75
7	Finished all over..........		4.00	4.80	6.40	8.75
8	Rough Body, Plated Trimmings....		3.75	4.65	6 25	8.00
9	" " " all over......		3.80	4.75	6.40	8.10
10	Finished and Plated all over		4.25	5.25	7.00	9.25

Order by these Numbers.

☞ If square ends are wanted, say whether right or left threads.

Genuine Jenkins' Patent Valves.

With Plumbago Disk.

Size..........	¼	⅜	½	¾	1	1¼	1½	2	2½	3 in.
Brass Globe or Angle..	1.10	1.25	1.60	2.20	2.80	4.00	5.50	8.00	15.75	22.00
Gate	......	......	1.50	2.00	2.85	4.00	5.00	7.50	14.00	20.00
Flanged Globe..........	......	......	......	......	......	9.00	11.00	16.50	25.00	34.00
Brass Safety or Angle.	......	......	3.75	4.50	5.00	7.50	9.25	14.00	....	....
Check..........	1.10	1.20	1.30	1.90	2.60	3.60	5 00	7.50	13.50	20.50

Jenkins' Patent Valves.

Plumbago Discs.—With Iron Bodies.

SIZE	1¼	1½	2	2½	3	3½	4	5	6	7	8 in.
Globe and Angle Valves, Screwed	3.85	5.00	7.25	11.00	16.00	......	......	......	......	......	......
" " " Flanged	......	......	8.50	13.00	18.00	......	......	......	......	......	......
" " " Screwed with Yoke	......	......	10.00	12.00	16.75	19.50	24.00	40.00	48.00	80.00	90.00
" " " Flanged	......	......	11.75	14.00	18.50	21.50	26.00	42.00	50.00	80.00	90.00
Check Valves, Screwed	......	......	......	10.50	14.00	17.00	20.00	30.00	40.00	......	......
" " Flanged	......	......	......	12.50	16.50	20.00	23.00	33.00	43.00	......	......
Cross Valves, Screwed	......	......	......	16.00	31.00	26.00	30.00	45.00	58.00	......	......
" " Flanged	......	......	......	19.00	24.00	29.00	33.00	48.00	62.00	......	......
Gate Valves, Screwed	......	......	8.00	12.00	15.00	18.00	21.00	30.00	36.00	50.00	62.00
" " Flanged	......	......	9.60	13.00	16.00	19.00	22.50	32.00	38.00	50.00	62.00
Safety Valves, Screwed	6.25	7.25	10.25	16.75	22.00	31.00	38.00	55.00	73.00	......	......
" " Flanged	......	......	12.25	19.00	25.50	34.00	41.50	62.00	80.00	......	......

Disks for all Styles Frink and Jenkins' Patent Valves.

SIZE	¼	⅜	½	¾	1	1¼	1½	2	2½	3	3½	4	5	6	7	8 in.
Price	.04	.05	.06	.07	.08	.12	.16	.24	.32	.40	.50	.60	.80	1.00	1.35	1.45

Chapman Composition Steam and Water Valves.

QUICK OPEN VALVE.

Diameter of Opening, inches,	1/4	3/8	1/2	3/4	1	1 1/4	1 1/2	2	2 1/2	3	3 1/2	4
Face to Face, Screw Ends	2 3/32	2 3/32	2 3/8	2 3/4	3 3/8	3 11/16	4 1/8	4 3/4	5 9/16	6 5/8	8 3/8	8 15/16
Face to Face, Flange Ends,	2 1/2	2 1/2	2 9/16	3	3 9/16	3 15/16	4 5/16	5 1/4	5 3/4	7	8 1/4	8 7/8
Diameter of Flanges,	2 1/2	2 1/2	3	3	4	4 1/2	5	6	7	7	8 1/2	9
Finished Weight, Screw Ends,	1 1/16 lbs	1 1/16 lbs	1 1/2 lbs	2 1/4 lbs	3 1/4 lbs	4 3/4 lbs	6 lbs.	9 3/4 lbs.	16 lbs.	22 1/2 lbs.	40 lbs	54 lbs.
Finished Weight, Flange Ends	1 1/2 lbs	1 1/2 lbs	2 1/4 lbs	3 1/4 lbs	4 3/4 lbs	7 1/4 lbs	9 lbs.	14 1/2 lbs.	22 lbs.	29 1/2 lbs.	50 lbs.	63 lbs.

PRICE LIST.

Screw End,	$1.20	$1.20	$1.30	$1.75	$2.25	$3.25	$4.25	$6.25	$11.50	$16.00	$30.00	$38.00
Flange End,	2.25	2.25	2.50	3.00	4.00	5.00	7.50	10.00	16.00	20.00	39.00	46.00
Sliding Stem and Lever, extra,	.75	.75	1.00	1.20	1.40	1.60	1.80	2.00	2.25	2.50	2.75	3.00

Chapman Iron Body Steam and Water Valves.

Bolted Top. Steam Babbitt Seats.

Diameter of Opening, inches	2½	3	3½	4	4½	5	6	7	8	10	12	14	15	16	18	20	24
Face to Face, Screw Ends	6⅝	7½	8⅜	9⅝	9¾	10¼	11⅜	12⅛	12½	......	......	......	......	......	......	......	......
Face to Face, Flange Ends	7⅞	8¼	8⅞	9⅜	10¼	9⅝	10⅞	11½	11⅞	13⅝	14⅝	15⅞	16⅝	18¾	20	21	24
Diameter of Flanges	7	7	8½	9	9½	10	11	12	13	16	18	21	22	23	25	27	31
Finished Weight, Screw Ends	28 lb.	40	55	71	99	120	166	225	270	......	......	......	......	......	......	......	......
Finished Weight, Flange Ends	40 lb.	49	71	82	90½	132	183	235	266	442	620	870	1,000	1,140	1,525	1,925	3,159

PRICE LIST.

Screw Ends	10.00	13.00	16.50	19.00	23.00	25.00	32.00	38.00	48.00	......	......	......	......	......	......	......	......
Flange Ends	10.00	13.00	17.00	18.50	22.00	24.00	31.00	37.00	45.00	64.00	86.00	......	......	......	......	......	......
Sliding Stem and Lever, Extra	2.25	2.50	2.75	3.00	......	......	......	......	......	......	......	......	......	......	......	......	......

Chapman Iron Body Water Gates.

Composition Mountings. Bell or Spigot Ends. Bolted Top. Without Gearing.

FOR STREET MAINS.

Diam. of Opening, inches,	2	3	4	5	6	7	8	10	12	14	15	16	18	20	24
End to end of Pipe when laid in Bell.	3 3/8	3 3/4	5 1/4	5 1/4	6	6	6 3/8	7 5/8	8	10	10	9 7/8	10 7/8	11 1/4	12 1/8
Diameter of Bell Socket,	3 1/8	4 5/8	5 3/4	6 7/8	7 7/8	8 7/8	10	12	14 3/8	16 1/4	17 1/2	18 1/2	20 5/8	22 3/4	26 3/4
Fin'd Weight, Bell Ends,	32 lbs.	55	116	135	195	245	290	439	600	843	950	1,080	1,475	1,700	2,750
Fin'd W'ght, Spigot Ends.	32 lbs.	67	108	127	185	240	296	470	701	946	1,022	1,229	....	....	......

PRICE LIST.

Bell or Spigot End,	$10.00	$15.00	$19.00	$25.00	$30.50	$36.00	$45.00	$62.00	$82.00	....	....	....	....	....	......

Kennedy's Brass Double Gate Valves.

Made of Superior Steam Metal. For Steam or Water.

These Valves are double faced, parallel and closely fitted, either end can be used for inlet or outlet. All parts are interchangeable. All our Valves are tested before shipment.

Diam. of opening,	3/8	1/2	3/4	1	1 1/4	1 1/2	2	2 1/2	3	3 1/2	4	5	6 in
Face to face, screwed ends,	......	2 1/4	2 5/8	3	3 1/4	3 1/2	4 1/8	4 3/4	4 7/8	5 3/4	6	....	....
Face to face, flanged ends,	....	....	3	3	3 1/2	4	4 3/4	5 1/2	6	6 1/4	7	....	9
Diam. of flanges,	....	3	3	4	4 1/2	5	6	7	7	7 1/2	9	10	11

Larger sizes to order. Prices on application.

(See Price list over.)

PRICE LIST.

(Kennedy's Brass Double Gate Valves, See preceding page.)

Screwed ends	$1.25	1.30	1.75	2.50	3.50	5.00	7.50	14.00	20.00	32.00	40.00	55.00	78.00
Flanged ends	2.50	2.75	3.50	4.50	5.50	7.50	12.00	18.00	25.00	40.00	48.00	66.00	94.00

Kennedy's Iron Body Double Gate Valves.

Brass Mounted.

Diameter of Opening	2	2½	3	3½	4	4½	5	6	7	8	10	12	14	15	16	18	20	24 in
Face to Face, Screw Ends	5⅛	6	6⅛	6¾	7⅛	7½	8	8¾	10	10¼	......	......	......	......	......	......	......	......
Face to Face, Flange Ends	5⅞	6⅛	7½	7½	8⅛	8½	9⅛	9⅞	9½	11	12⅛	13⅛	13¾	......	14	14⅞	15½	17
Diameter of Flanges	6	7	8	8½	9	9½	10	11	12	13	16	18	21	......	23	25	27	31

PRICE LIST.

	2	2½	3	3½	4	4½	5	6	7	8	10	12	14	15	16	18	20	24
Screw Ends	8.50	12.00	15.00	18.00	20.00	22.50	25.00	30.00	40.00	50.00	65.00	90.00	......	......	......	......	......	......
Flange Ends	9.00	12.50	15.50	19.00	21.00	24.00	27.00	32.00	40.00	50.00	65.00	90.00	......	......	......	......	......	......
Sliding Stem and Lever, Screw Ends	11.00	14.75	18.00	21.50	24.00	27.00	30.00	36.00	47.00	58.00	......	......	......	......	......	......	......	......
Sliding Stem and Lever Flange Ends	11.50	15.25	18.50	22.50	25.00	28.50	32.00	38.00	47.00	58.00	......	......	......	......	......	......	......	......

Prices on larger sizes, up to 60 inches, given on application.

Kennedy's Iron Body Water Gates, Brass Mounted, Hub or Spigot Ends.

To bear medium pressure on either side of gates.

These Valves are operated by a two-inch square nut on spindle unless otherwise ordered.

The bodies, caps, nuts, stuffing boxes and glands are made of cast iron; the gates are also cast iron faced with brass. The seats are of brass, firmly held to the body according to the most approved practice. The Stems are large to prevent twisting and of solid gun metal composition, and are all interchangeable.

These Valves are specially constructed for street mains and are extra strong to withstand rough usage.

SIZE	3	4	5	6	7	8	10	12	14	16	18	20	24
End to End of pipe when laid in Hub	3¼	4	5	5	4¾	5¾	5¾	7¼	8	6½	6	9	10
Diameter of Hub Socket	4⅝	5¾	6⅞	7⅞	8⅞	10	12	14¼	16¼	18½	20½	22¾	26¼

PRICE LIST.

	3	4	5	6	7	8	10	12	14	16	18	20	24
Hub or Spigot End	$15.00	20.00	25.00	30.00	40.00	50.00	65.00	90.00	....	....	....	....	....

When ordering state if valves should open by turning to the left or to the right; when not mentioned, we send valves which open by turning to the **left.**

These Valves have Brass Stuffing Boxes on all sizes up to 6 inches.

Rensselaer Double Gate Brass Valves.

Best Steam Metal. Gland and Follower Packing Box.

Diameter of Opening, inches,.......	½	¾	1	1¼	1½	2	2½	3	3½	4	5	6
Face to Face, Screw Ends,.......	2¼	2½	2⅞	3⅜	3¾	4⅛	4⅞	5	......	7¼	......	8¼
Face to Face, Flange Ends,.......	......	......	......	......	......	4⅝	5⅜	6¼	......	7	......	9
Diam. of Standard Flanges,.......	......	......	......	......	......	6	6½	7	......	9	......	11

PRICE LIST.

Screw End,.......	$1.25	$1.65	$2.15	$3.15	$4.25	$6.25	$11.50	$16.00	$21.00	$35.00	$52.00	$78.00
Flange End,...........................	......	......	......	......	......	11.50	18.00	22.00	31.00	43.00	64.00	90.00

In packing these Valves see that the Follower Gland Ring is *above the packing.* Never allow face or seat to become daubed with lead.

Rensselaer Double Gate, Iron Body, Brass Mounted Valves.

Screwed or Flanged Ends. To Bear Heavy Pressure Either Side of Gate.

Diam. of Openings, inches,..........	2	2½	3	3½	4	5	6	7	8	10	12	14	16	18	20	24	30	36	42	48	60
Face to Face, Flange ends,...	5⅝	6¼	6⅞	9½	8¼	10¾	11¼	11½	11¾	12⅞	13	15¾	16	17	18	21	Depends upon size by-pass required.				
Diam. of Standard Flanges,...	6½	7	8	8½	9	10	11	12	14	16	18	21	23	25	27	31	38	44	51	58	71

PRICE LIST.

Screw End......... ...	$7.00	$10.50	$13.00	$16.50	$18.00	$25.00	$31.00	....	....	....	....	...	...	...	...	...	...	...	...	...	...
Flange Ends....	7.25	10.75	13.50	17.00	18.50	24.50	30.00	37.00	43.50	60.00	78.00	...	...	...	...	...	...	...	...	...	...

(For list on Rensselaer Water Gates, See page 44.)

LUDLOW'S PATENT SLIDING STOP VALVES.

Double Gate Brass Valves.—Gland in Packing Box.

SIZE.	Screw Socket.	Flange.	Diameter Standard Flange.	Face to Face of Screw Socket.	Face to Face of Flanges	Extra for Slide Stem and Lever. Subject to Discount
Inches.			Inches	Inches.	Inches.	
½	1.40	....	....	2¼	....	80
¾	1.80	....	....	2½	....	80
1	2.35	....	4	2⅞	3	80
1¼	3.40	5.70	4⅜	3⅜	3⅜	1.00
1½	4.40	7.40	5	3¾	3⅞	1.00
1¾	......	....	5¾	....	4 7/16	1.00
2	6.25	11.00	6	4⅛	4⅝	1.25
2½	13.75	18.75	6½	4 13/16	5⅜	1.75
3	15.50	21.50	7	5	6¼	2.00
3½	23.50	30.50	7½	5⅝	6⅞	2.00
4	34.00	43.00	9	6¼	7 1/16	2.00
4½	45.00	55.00	9½	7	7¾	2.25
5	52.00	64.00	10	7¼	8 13/16	2.25
6	76 00	88.00	11	7½	9	2.25
8	....	....	....	....	....	....
9	....	....	....	....	....	....
10	....	....	....	....	....	....
12	....	....	....	....	....	....
14	....	....	....	....	....	....
16	....	....	....	....	....	....
18	....	....	....	....	....	....
20	....	....	....	....	....	....
24	....	....	....	....	....	....
27	....	....	....	....	....	....
	....	....	....	....	....	....

LUDLOW'S IRON BODY VALVES, BRASS MOUNTED, DOUBLE GATE.

To Bear Heavy Pressure Either Side of Gate.

LIST PRICES BRASS MOUNTED.						MEASUREMENTS. BRASS MOUNTED OR ALL IRON.				
SIZE.	Screw Socket.	Flange.	Hub.	Spigot.	Extra for Slide Stem and Lever. Subject to Discount	Diameter Standard Flange.	Face to Face of Flang's	Face to Face of Screw Socket.	End to End of Hubs.	Depth of Hubs.
Inch.						Inches.	Inches	Inches	Inches.	Inch.
1	$ 5.00	......	......	......	......	......	......	3¾	......	......
1¼	5.50	......	.....	......	......	......	......	3¾	......	......
1½	6.00	$ 6.25	......	......	$1.00	5½	5½	4	......	......
2	7.00	7.50	$ 7.00	$ 7.25	1.25	6½	5⅝	4⅞	7	2¼
2½	10.25	10 75	10.00	10.25	1.75	7	6⅛	5⅜	7¾	2¼
3	12.25	13.25	14.50	15.00	2.00	8	6⅞	5½	8⅝	2½
3½	16.50	17.50	16.00	16.50	2.00	8½	8¼	7	9⅜	2⅝
4	18.00	18.50	17.00	17.50	2.00	9	8¼	7¼	9⅜	2¾
4½	23.00	23.50	22.00	22.50	2.25	9½	9¾	9½	......	......
5	25.00	25.50	24.00	24.50	2.25	10	10¾	11	12	3½
6	30.50	31.00	28.00	28.75	2.25	11	11⅛	11¼	12½	3⅞
7	38.00	38.00	37.00	38.00	2.25	12	11¼	12	13⅞	......
8	45.00	43.50	42.00	43.25	2.25	13	11	12⅝	14⅛	4
10	64.00	64.50	60.00	62.50	3.25	16	13¾	13⅝	14½	4
12	82.50	80.00	76.00	79.50	4 00	18	14⅝	13⅝	15	4

In ordering Valves, please give full details, stating whether Single or Double Gate, Brass Mounted or otherwise, Screw Socket, Flange, Hub or Spigot, Hand-wheel or Nut; also, for what the Valve is to be used, whether to open by turning to the right or to the left, and if not to stand upright whether to lie on edge, see page 16, figure 7, or on flat side.

Rensselaer Iron Body Water Gates.

Double Gate, Brass Mounted, Without Gearing, Bell or Spigot Ends.

Diam. of Opening, inches	2	2½	3	3½	4	5	6	7	8	10	12	14	16	18	20	24	30	36	42	48	60
End to end of Pipe when laid in Bell	2¾	......	3⅜	......	4¾	5¾	6	6	6½	7¼	7½	7¾	8¾	9½	10¼	11	Depends upon size by-pass required.				
Depth of Bell Socket	2⅛	......	2¾	......	3⅛	3½	4	4	4½	4	4¼	4¼	4¼	4¾	5	5	5	5	5	5	5

PRICE LIST.

Bell or Spigot Ends 7.25 10.75 14.50 16.00 17.50 24.00 28.00 36.00 42.00 58.00 76.00

Valves 16 in. and upwards may be furnished with gearing to stand upright, or with bevel gearing and anti-friction rolls to lie on side.

Brass and Iron Body Asbestos Disc Globe and Angle Valves.

Sizes...Inches	¼	⅜	½	¾	1	1¼	1½	2	2½	3
Globe and Angle Valves, Brass	$ 1.10	1.25	1.60	2.20	2.80	4.00	5.50	8.00	15.75	22.00
Patent Asbestos Discs	.06	.07	.09	.10	.12	.18	.25	.36	.48	.60

Sizes...Inches	2	2½	3	3½	4	5	6	7	8	10
Globe and Angle Valves, I. B., Brass Hub, Sc.	$ 7.25	11.00	16.00	....	....	....	....	....	....	....
Globe and Angle Valves, I. B., Brass Hub, Flgd.	8.50	13.00	18.00	....	....	....	....	....	....	....
Globe and Angle Valves, I. B., Sc., with Yoke	10.00	12.00	16.75	19.50	24.00	40.00	48.00	80.00	90.00	130.00
Globe and Angle Valves, I. B., Fl., with Yoke	11.75	14.00	18.50	21.50	26.00	42.00	50.00	80.00	90.00	130.00
Patent Asbestos Discs	.36	.48	.60	.75	.90	1.20	1.50	1.80	2.10	2.70

Asbestos Seat Brass Gate Valves, Screwed and Flanged.

SIZE	3/8	1/2	3/4	1	1 1/4	1 1/2	2	2 1/2	3
Screwed	$1.25	1.60	2.20	2.80	4.00	5.50	8.00	15.75	22.00
Flanged	......	......	5.00	6 00	9.00	11 00	16.50	25.00	34.00
Seat Ring	.14	.18	.20	.24	.36	.50	.72	.96	1.20

Asbestos Seat Iron Body Gate Valves, With Screwed Bonnet.

SIZES......Inches.	1/2	3/4	1	1 1/4	1 1/2	2	2 1/2	3	3 1/2	4
Screwed	$1.60	2.20	2.80	4.00	5 50	8.00	12.00	15.00	18.00	21.00
Flanged	1.85	2.50	3 25	4.50	6.00	8.00	12.00	15.00	18.00	21.00
Seat Ring	.18	.20	.24	.36	.50	.72	.96	1.20	1.50	1.80

Asbestos Seat Iron Body Gate Valves, With Bolted Bonnet.

SIZES	2	2 1/2	3	3 1/2	4	4 1/2	5	6	7	8	10	12
Screwed	$ 8 00	12.00	15.00,	18.00	21.00	25.00	30.00	36.00	50.00	62.00	....	....
Flanged	8.00	12.00	15.00	18.00	21.00	25.00	30.00	36.00	50 00	62.00	85 00	120.00
Seat Ring	.72	.96	1.20,	1.50	1.80	2.00	2.40	3.00	3.60	4.20	5.40	6 00

Asbestos Seat Iron Body Gate Valves, with Bell Ends.

Sizes	2	3	4	5	6	8	10	12
Price	$ 8.00	15.00	21.00	30.00	36.00	62.00	85.00	120.00
Seat Ring	.72	1.20	1.80	2.40	3.00	4.20	5.40	6.00

Brass Straightway, Swinging, Check, Stop, Back Pressure and Lock Check Valves.

Sizes	1/4	3/8	1/2	3/4	1	1¼	1½	2	2½	3
Brass Check Valves	$1.25	1.25	1.30	1.75	2.25	3.25	4.25	6.35	11.50	16.00
Brass Check Valves, Angle	....	....	1.30	1.75	2.25	3.25	4.25	6.25	....	....
Brass Stop Valves	1.25	1.25	1.30	1.75	2.25	3.25	4.25	6.25	11.50	16.00
Brass Back Pressure Valves	....	....	....	....	....	....	4.25	6.25	11.50	16.00
Brass Lock Check Valves	1.25	1.25	1.30	1.75	2.25	3.25	4.25	6.25	11.50	16.00

Iron Body Straightway, Swinging Check and Back Pressure Valves.

Size	2	2½	3	3½	4	5	6	7	8	10	12	14	16 in.
I. B. Check Valves, Fl., Sc. and Bell Ends	$6 25	10.00	12.00	16.00	18.00	25.00	32.00	41.00	50.00	65.00	95.00	....	....
I. B. Back Pressure Valves, Sc. and Fl	6.25	10.00	12.00	16.00	18.00	25.00	32.00	41.00	50.00	65.00	95.00	....	....

Iron Cocks, Asbestos Packed, Screwed, Flanged and Gland Ends.

SIZES	¼	⅜	½	¾	1	1¼	1½	2	2½	3	3½	4	5	6 in.
Regular Screwed and Flanged	$1.30	1.45	1.60	2.10	2.50	3.50	4.75	7.00	12.00	18.00	27.00	30.00	45.00	60.00
Stop and Waste, Screwed and Flanged	1.30	1.45	1.60	2.10	2.50	3.50	4.75	7.00	12.00	18.00	....	....	....	....
Extra Heavy, Screwed and Flanged	1.45	1.60	2.10	2.50	3.50	4.75	7.00	12.00	18.00	27.00	30.00	45.00	....	....
For Superheated Steam, Screwed and Flanged	1.45	1.60	2.10	2.50	3.50	4.75	7.00	12.00	18.00	27.00	30.00	45.00	60.00	....
With Worm and Gear, Screwed and Flanged	....	....	....	....	....	....	....	18.00	25.00	30.00	35.00	45.00	65.00	75.00
For Ammonia, Screwed	1.30	1.45	1.60	2.10	2.50	3.50	4.75	7.00	12.00	18.00	27.00	30.00	45.00	....
For Ammonia, Gland	1.45	1.60	2.10	2.50	3.50	4.75	7.00	12.00	18.00	27.00	....	....	....	....
Three Way Cocks, "A. & B." Pat., Screwed	3.00	3.25	3.75	5.25	7.25	10.00	18.00	25.00	30.00	35.00	50.00	....	....	....
Brass Cocks, Screwed	2.00	2.25	2.50	3.15	4.20	6.00	7.75	12.00	20.00	28.00	....	....	....	....
Iron Cock Wrenches	.10	.10	.10	.20	.20	.30	.40	.50	1.00	1.50	1.60	1.75	....	....

Dimensions of Iron Body Asbestos Valves.

SIZES	2	2½	3	3½	4	5	6	7	8	10	12 in.
Globe Screwed, End to End	6	7⅛	8⅛	9⅛	10⅜	11¼	13¾	14½	16¼	20	23¾
Globe Flanged, Face to Face	6½	8	8½	10⅛	10¾	12	15½	15⅝	17	20	23¾
Globe Flanged, Size of Flanges	6½	7	8	9	10	11	12	13	14	16	19
Angle Screwed, Centre to Face	3	4¼	4⅛	4½	5¼	5⅞	7	7¼	8⅛	10	11⅞
Angle Flanged, Centre to Face	4⅛	4⅝	4¼	5 1/16	5½	6¼	7⅜	7 13/16	8½	10	11⅞
Angle Flanged, Size of Flanges	6½	7	8	9	10	11	12	13	14	16	19
Stop Screwed, End to End	5⅞	7⅜	8	9¾	10	12	13½	14½	15	19⅜	22½
Stop Flanged, Face to Face	8¾	8¼	8	9⅝	10½	11⅞	13⅜	14½	15⅜	19¾	22¼
Stop Flanged, Size of Flanges	6	7	7	8½	9	10	11	12	13	16	18
Check Screwed, End to End	5	7⅜	8	9¾	10	12	13½	14½	15	19⅜	22¼
Check Flanged, Face to Face	8⅜	8¼	8	9⅝	10½	11⅞	13⅜	14½	15⅜	19¾	22½
Check Flanged, Size of Flanges	6	7	7	8½	9	10	11	12	13	16	18

Dimensions of Asbestos Packed Brass and Iron Cocks.

SIZES	1/4	3/8	1/2	3/4	1	1 1/4	1 1/2	2	2 1/2	3	3 1/2	4	5	6 in.
Brass Screwed, End to End	2 3/4	2 7/8	2 7/8	3 3/8	4	4 1/2	5	6	.....	.....	.....	.....	.....	.....
Brass Screwed, Top of Plug to Bottom	3 1/8	3 5/8	3 5/8	4 3/8	5	5 7/8	6 7/8	8	.....	.....	.....	.....	.....	.....
Iron Screwed, End to End	2 3/4	2 7/8	2 7/8	3 3/8	4	4 9/16	5 1/8	6 1/2	7	8 3/4	9 1/2	10 7/8	12 5/8	14 3/4
Iron Screwed, Top of Plug to Bottom	3 1/8	3 5/8	3 5/8	4 1/2	5 1/4	6 1/8	7 1/8	8 1/3	9 3/4	10 3/4	12	13 1/3	16 1/6	19
Iron Flanged, Face to Face	.....	.....	.....	.....	.....	.....	5 7/8	7 1/8	8 1/4	9 1/4	10 1/4	11 1/2	13 3/4	16
Iron Flanged, Top of Plug to Bottom	.....	.....	.....	.....	.....	.....	7 1/8	8 1/8	9 3/4	10 3/4	12	13 1/4	16 1/2	19
Iron Flanged, Size of Flanges	.....	.....	.....	.....	.....	.....	6	7	7	7	8 1/2	9	10	11

Dimensions of Brass and Iron Body Asbestos Gate Valves.

Brass Gate Valves.

SIZES	1/2	3/4	1	1 1/4	1 1/2	2	2 1/2	3	3 1/2	4 in.
Distance E to E, Screwed	2 1/2	2 7/8	3 5/16	3 3/4	4 1/8	4 7/8	6 5/8	7 1/2	8 3/8	8 7/8
Distance F to F, Flanged	2 13/16	3 3/16	3 11/16	4 1/8	4 5/16	5 3/8	6 3/4	8 1/8	9 1/8	10
Diameter of Flanges	3	3	4	4 1/2	5	6	7	8	8 1/2	9

IRON BODY GATE VALVES.

SIZES	2	2 1/2	3	3 1/2	4	4 1/2	5	6	7	8	10	12	14	16	18	20	24 in
Distance E to E, Screwed	5 3/4	6 5/8	7 1/2	8 3/8	8 7/8	9 1/2	9 1/2	10	10 7/8	11 1/2	.....	.....	.....	.....	.....	.....	.....
Distance F to F, Flanged	7 1/4	7 7/8	8 1/4	8 7/8	9 1/2	10 1/2	10 1/2	10 7/8	11 7/8	12 1/6	13 3/4	16 3/4	18	19	20	22	24
Height over all of Stationary Spindle	11 5/16	14 1/2	15 1/8	18	19 3/4	21	23	25 1/4	28 1/4	31 3/8	36 1/4	42 5/8	51 3/8	56	60 1/2	67 1/2	80
Height over all of Rising Spindle	.....	19 3/4	24 3/4	22	29 1/6	25 3/4	35	40	45 1/4	50 5/8	60	.....	.....	.....	.....	.....	.....
Height over all of Indicator Spindle	.....	.....	16 3/4	20	21 1/4	22 1/2	26	28 1/3	32 1/2	34 7/8	39 7/8	.....	.....	.....	.....	.....	.....
Diameter of Flanges	6 1/2	7	8	8 1/2	9	9 1/2	10	11	12	13	16	18	21	23	25	27	31

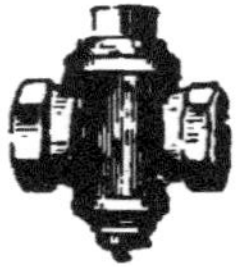

Steam Cocks.

Square or Flat Head.

SIZE,	1/8	1/4	3/8	1/2	3/4	1	1¼	1½	2	2½	3	3½	4 in.
Brass,	.70	.70	.75	1.10	1.50	2.25	3.75	4.80	7.25	14.00	20.00	....	....
" Extra heavy,	.70	.70	.75	1.10	1.50	2.25	3.75	4.80	7.25	14.00	20.00	36.00	50.00
" Three-way,	.....	.....	.....	1.75	2.40	3.60	5.75	7.30	10.40	18.50	26.75	45.00	65.00

Service Cocks.

Flat or Square Heads.

SIZE	¼	3/8	½	¾	1	1¼	1½	2	2½	3	3½	4 in.
Female	.55	.65	.75	1.00	1.40	2.20	3.00	5.00	10.00	15.60		
Male and Female, or Extra Heavy	.65	.75	.85	1.20	1.70	2.60	3.60	5.75	11.50	17.00		

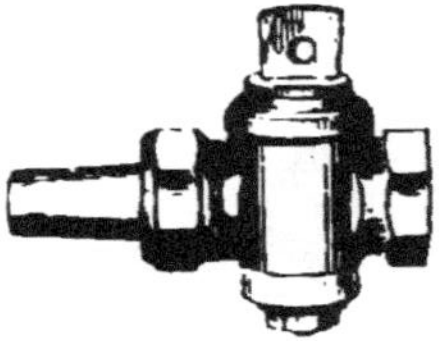

Metre Cocks.

SIZE	3/8	½	¾	1	1¼	1½	2	2½	3 in.
Without Union,	.70	.85	1.20	1.70	2.60	3.60	5.75	11.50	17.00
With Union,	......	1.00	1.30	2.00	3.00	4.25	6.75	......	....

Air Cocks.

SIZE	1/8	1/4	3/8	1/2 in.
Single Thread	.40	.45	.50	.60
Double "	.45	.50	.90	1.00
Bibb Nozzle, Single Thread	.65	.70	.75	.85
" " Double "	.70	.90	1.00	1.25

Compression Radiator Air Valves.

No. 1.

No. 2.

No. 3.

No. 4.

SIZE	1/8	1/4 in.
No. 1, Plain	.45	.50
" 1, Nickle Plated	.50	.55
No. 2, Plain	.35	.40
" 2, Nickle Plated	.40	.45
No. 3 W. Wheel, Plain	.65	.70
" 3 " Plated	.70	.75
No. 4, with Key, Plain	.30	.35
" 4 " " Plated	.35	.40

Keys for above, extra plain 4c. Plated, 6c. net.

VAN AUKEN'S

Duplex Automatic Air Valves

FOR STEAM RADIATORS.

These Valves will not permit steam or water to escape from the Radiator.

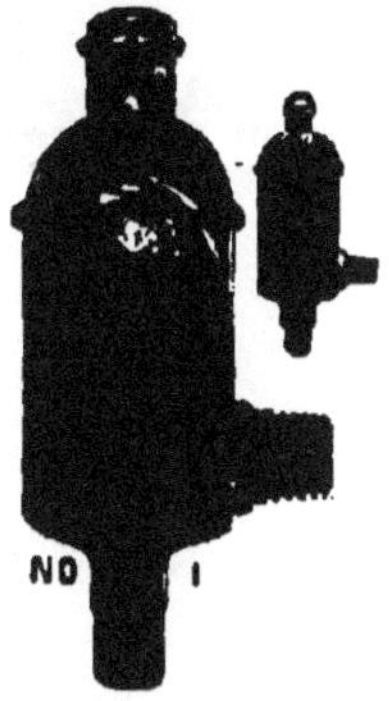

STANDARD

Price, - - $1.15.

Order by Number only, and state whether Valves are to be used for High or Low Pressure.

Pronounced by the Trade generally, and the users, as the only perfect Automatic Air Valves on the market.

Cylinder Cocks.

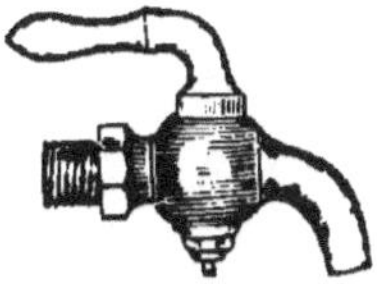

SIZE, Chased Iron Pipe......	¼	⅜	½	¾ in.
Price Tee Handles each,..........	.85	.95	1.25	2.25
" Lever " "	1.00	1.10	1.50	2.50

No. 7.

SIZE, Chased Iron Pipe,..	¼	⅜	½	¾ in.
Price, Tee Handles, each,......	.95	1.05	1.40	2.40
" Lever " "	1.00	1.20	1.65	2.65

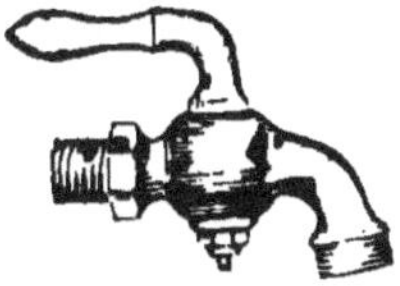

No. 8.

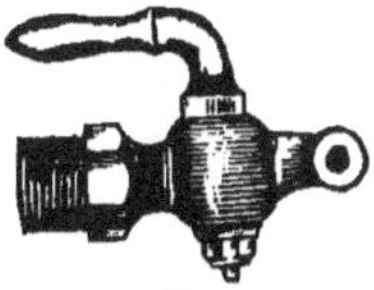

SIZE, Chased Iron Pipe......	¼	⅜	½	¾ in.
Price, Tee Handles, each,.........	.90	1.00	1.30	
" Lever " "	1.05	1.15	1.55	......

No. 9.

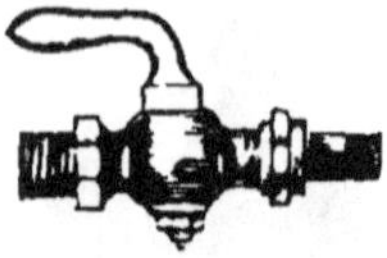

No. 10.

SIZE, Chased Iron Pipe	3/8	1/2	3/4 in.
Price, Lever Handles, each	1.75	2.25	3.00

Valve Guage Cocks.

Lever Handle.

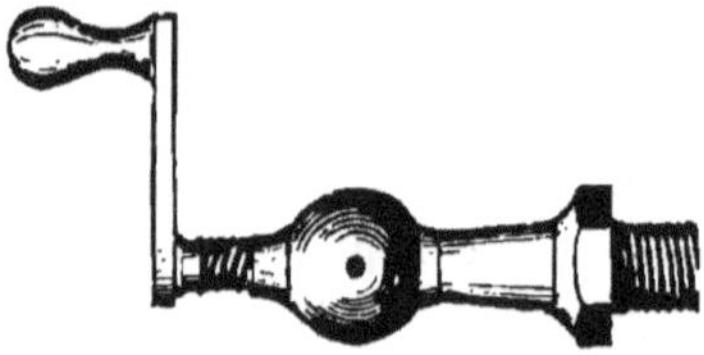

SIZE	3/8	1/2	3/4 in.
	.85	1.00	1.25

Compression Guage Cocks.

Wood Handles.

SIZE, Chased Iron Pipe		3/8	1/2	3/4 in.
No. 1,	Price	.80	.85	......
No. 2,	"	......	1.00	1.10
No. 3, Large Pattern,	"	......	1.50	1.60
No. 4 with Stuffing Box,	"	1.10	1 15	1.25
No. 5, " " "	"	......	1.65	1 75

Register's Patent Guage Cocks.

With Iron Balls, Cut or Blank, 3/8, 1/2, 3/4 in., $1.00

Steam Guage Cocks.

SIZE	1/4 in.
Male and Female	.75
Female	.75

Plain Oil Cups.

Diameter of body,	3/4	7/8	1	1 1/8	1 1/4	1 1/2	1 3/4	2	2 1/4	2 1/2	3 in.
Size, Chased Iron Pipe,	1/8	1/8	1/4	1/4	1/4	3/8	3/8	3/8	1/2	1/2	3/4
Price,	.30	.35	.40	.50	.60	.90	1.25	1.75	2.25	2.75	4.50

New Pattern Locomotive Cups.

Diameter of body	1	1 1/4	1 1/2	2 in.
Size, Chased Iron Pipe	1/4	1/4	3/8	3/8
Price	.50	.75	1.00	2.00

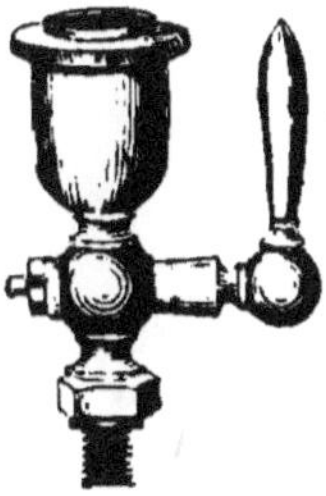

Lever Handle Oil Cups.

Diameter of body	1 1/8	1 1/4	1 1/2	2	2 1/4 in.
Size, Chased Iron Pipe	1/4	1/4	3/8	1/2	1/2
Price	1.35	1.60	2.20	3.2[illegible]	4.00

T Handle Oil Cups.

Diameter of body	1	1 1/4	1 1/2	2	2 1/4 in.
Size, Chased Iron Pipe	1/4	1/4	3/8	1/2	1/2
Price	1.00	1.50	2.00	3.00	3.75

D

Globe Oil Cups.

Diameteter...............	1½	2	2¼	2½	3	3½	4 in.
Number....................	00	0	1	2	3	4	5
Size.......	3/8	3/8	½	½	½	3/4	3/4
Price...........	3.00	3.75	4.50	5.50	6.50	9.00	15.00

Common Lubricator.

Wood Handles.

Number...	1	2	3	4	5	6	7	8	9	10	11
Diameter.	3/4	1	1¼	1½	1¾	2	2¼	2½	3	3½	4 in.
Size.........	¼	¼	3/8	3/8	½	½	½	½	¾	¾	¾ in.
Price.......	1.75	2.00	2.20	2.40	2.60	2.90	3.25	3.75	4.75	7.00	10.00

Nos. 03 and 6.

Felthousen's Patent Oil Pumps.

To go on top of Cyliuder or Steam Pipe.

Description.	Size.	With Screw Top.	With Strainer Top.	Extra Str. iner Top.
Price No. 03 Finished, each,..	2 x 2	3.50	3.65	.40
" " 6 " "	2¾ x 2¾	5.00	5.25	.40

Hand Cylinder Oil Pumps.

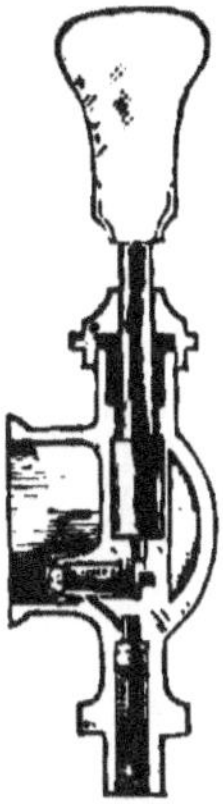

	Size.	With Screw Top.	With Strainer Top.	Extra Strainer Tops.
No. 3 Finished, each............	2 x 2	3.50	3.65	.40
" 04 " "	2¾ x 2¾	5.00	5.25	.60
" 4 " "	3½ x 3½	7.50	7.80	.75
" 5 " "	7 x 4½	12.00	12.00	......

Lever Handle Oil Pumps.

Price, each,..................................$11.00

Pioneer Oil Cups.

Number..........................	000	00	.0	1	1½	2	3	4	5	6
Outside diameter of Glass..........	1	1⅛	1¼	1½	1¾	2	2¼	2½	3	3½
Height....................................	⅞	1	1⅛	1⅜	1⅝	1⅞	2⅛	2⅜	3	4
Capacity, ounces..........................	¼	½	⅝	1	1½	2½	4	5	10	18
Pipe thread...............................	⅛	⅛	⅛	¼	¼	⅜	⅜	⅜	½	½
Finished brass, each......................	.70	.75	.80	1.00	1.25	1.50	1.90	2.40	3.10	4.00
Nickel, " "	.80	.85	.95	1.20	1.50	1.75	2.20	2.75	3.50	4.50
Extra glasses, net.	.05	.06	.08	.10	.10	.12	.15	.25	.35	.65

Royal Oil Cups.

Sight Feed.

Number..............	00	0	1	1½	2	3	4	5	6
Outside dia. of Glass	1⅛	1¼	1½	1¾	2	2¼	2½	3	3½
Height.	1	1⅛	1⅜	1⅝	1⅞	2⅛	2⅜	3	4
Capacity, ounces....	½	⅝	1	1½	2½	4	5	10	18
Pipe thread..........	⅛	⅛	¼	¼	⅜	⅜	⅜	½	½
Finished brass, each	1.10	1.25	1.50	1.75	2.10	2.55	3.15	3.90	4.80
Nickel " "	1.20	1.40	1.70	2.00	2.35	2.85	3.50	4.30	5.30
Extra Glasses, net..	.06	.08	.10	.10	.12	15	.25	.35	.65

Crank Pin Oiler.

Automatic.

Number	0	1	1 1/2	2	3	4
Outside diameter of Glass	1 1/4	1 1/2	1 3/4	2	2 1/4	2 1/2
Height of glass	1 1/8	3/8	1 5/8	1 7/8	2 1/8	2 3/8
Pipe thread	1 1/8	1/4	1/4	3/8	3/8	3/8
Price, each	1.10	1.50	2.00	2.50	3.00	4.00
Extra Glasses, net, each	.08	.10	.10	.12	.15	.25

Shafting Oilers.

No. 9.

No. 10.

No. 10, W. B.

No.	9	9 W. B.	10	10 W. B.
Diameter	2 1/2	1 3/4	2 1/4	2
Height	3 1/4	3 1/4	4 1/4	4 1/4
Capacity, ounces	1 1/2	1 1/2	3 1/2	3
Price per. doz	$4.50	4.50	4.50	4.50

Detroit Sight Feed Lubricators.

SIZE.	Plain Brass.	Nickel Plated	Suitable for Engine with Diameter of Cylinder as follows:
Third Pint	....	....	....
Half Pint	$22.00	$25.00	Up to 10 inches.
Pint	30.00	35.00	10 to 18 inches.
Quart	45.00	50.00	18 to 30 inches.
Half Gallon.	60.00	65.00	30 and over.

We also keep in stock the above Lubricators with Single Connections.

Steam Engine Indicators of any kind furnished to order.

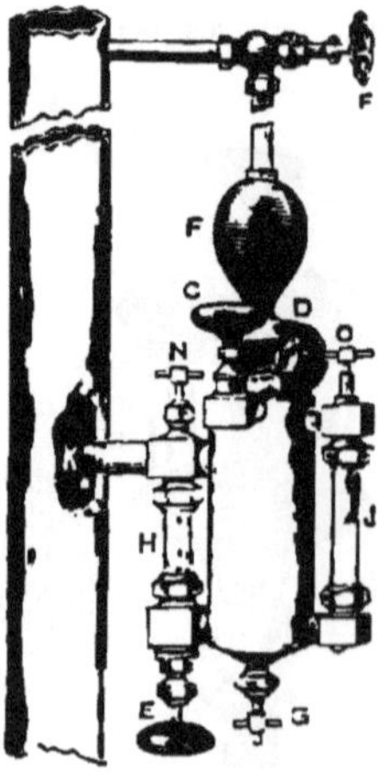

Hall's Pneumatic Sight Feed Oilers.

For Shafting, Engine Bearings, &c.

	Diameter.	Height.	Capacity, Ounces.	Price Each.	Price per dozen.
No. 1..........	1⅝ inch.	3 inch.	¾	$1.50	$ 16.00
No. 2..........	2 "	3¼ "	1¼	1.75	20.00
No. 3..........	2¼ "	3½ "	2	2.25	24.00
No. 4..........	2½ "	3¾ "	3	2.50	28.00
No. 6..........	3 "	4½ "	5	3.00	34 00
No. 8..........	3½ "	5 "	8	3.50	40.00

The Metropolitan Automatic Injector.

Operated Entirely by One Handle..

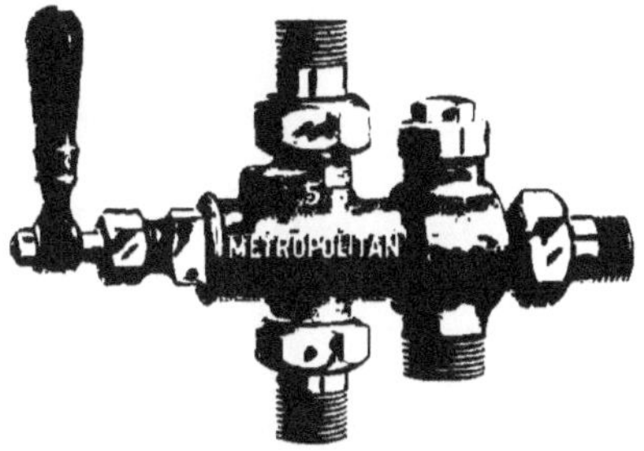

Price List of Metropolitan Injectors.

Size.	Price.	Size of Pipe Connection.	Gallons Per hour 65 lbs. Pressure.	Horse Power.
2....................	$ 15.50	⅜	40	4 to 6
3....................	16 00	⅜	60	6 to 8
3½....................	18.00	½	90	8 to 12
4....................	20.00	½	120	12 to 16
5....................	25.00	¾	220	16 to 28
6....................	30.00	¾	300	28 to 40
7....................	40.00	1	420	40 to 57
8....................	45.09	1	540	57 to 72
9....................	55.00	1¼	720	72 to 93
10....................	60.00	1¼	900	93 to 120
11....................	75.00	1½	1260	120 to 168
12....................	90.00	1½	1740	168 to 232
13....................	110.00	2	2240	232 to 298
14....................	125.00	2	2820	298 to 382

The Hancock Inspirator.

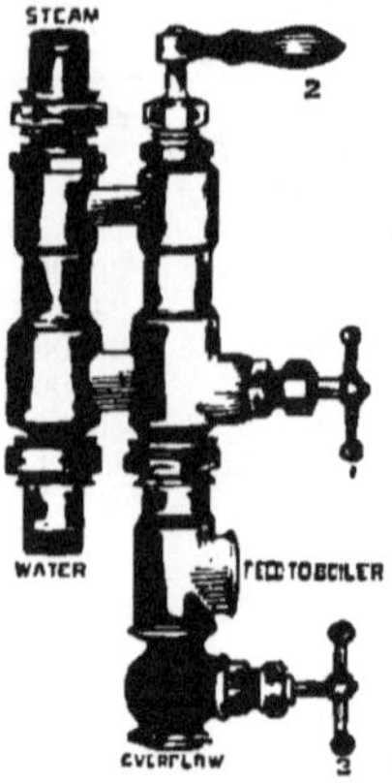

No. of Inspirator.	Size of Connections.		Gallons per hour, 60 lbs. Pressure.	Price.
	Suction and Feed.	Steam.		
No. 7½	⅜	⅜	60	$ 16.00
" 8¾	½	⅜	85	18.00
" 10	½	⅜	120	20.00
" 12½	¾	½	220	25.00
" 15	¾	½	300	30.00
" 17½	1	¾	360	40.00
" 20	1	¾	540	45.00
" 22½	1¼	1	700	55.00
" 25	1¼	1	900	60.00
" 30	1½	1¼	1260	75.00
" 35	1½	1¼	1740	90.00
" 40	2	1½	2230	110.00
" 45	2	1½	2820	125.00
" 50	2½	2	3480	150.00

Friedmann's Patent Injectors—For Feeding Steam Boilers.

Size of Injector	Minimum Inside Diameter of Pipe in Inches.	Gallons per Hour. 80 lbs. Pressure.	PRICE Class C, Non-Lifting.	PRICE Class D, Lifting.
No. 2	½	110	17.00	19.00
" 3	¾	180	27.00	32.00
" 4	1	320	40.00	45.00
" 5	1¼	500	50.00	55.00
" 6	1¼	720	60.00	65.00
" 7	1½	965	75.00	80.00
" 8	1½	1280	90.00	100.00
" 9	2	1620	110.00	120.00
" 10	2	2000	130.00	140.00
" 12	2½	2880	160.00	180.00

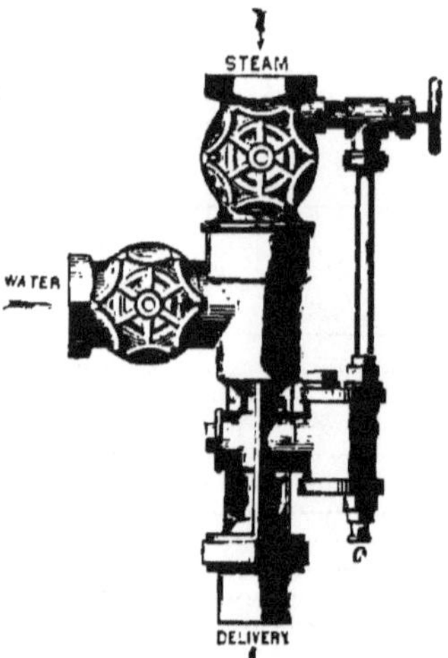

Class D.

Ejectors or Water Elevators.

For Raising Water and Conveying Liquids.

NUMBER.	000	00	0	1	2	3	4
Delivery per hour in Gallons at 45 lbs. Steam Pressure...	250	500	900	1200	2000	3000	5000
Diameter of Steam Pipe in inches,	3/8	1/2	3/4	3/4	1 1/4	1 1/4	1 1/2
Diameter of Delivery Pipe in ins.,	1/2	3/4	1	1 1/4	1 1/2	2	2 1/2
Diameter of Suction Pipe in ins.,	3/4	1	1 1/4	1 1/2	2	2 1/2	2 1/2
Boiler Capacity, Horse Power....	3 to 4	3 to 4	3 to 4	5 to 6	7 to 8	10 to 15	25
Price....	$8.00	14.00	20 00	30.00	50.00	75.00	100.00

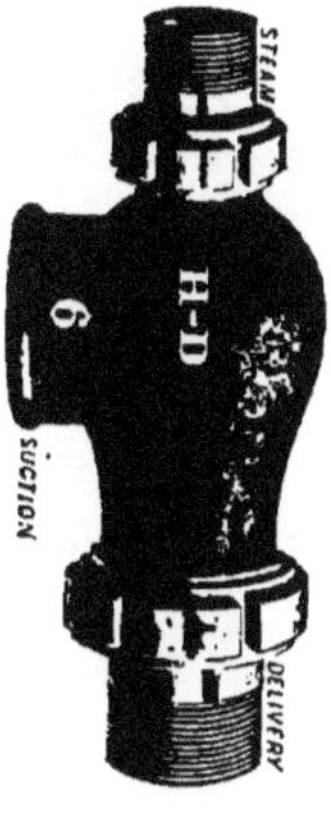

The H-D Ejector or Jet Pump.

Price List and Capacity.

SIZE.	PRICE.	PIPE CONNECTIONS.		Capacity per Hour.	STRAINERS.
		Steam	Suction and Delivery.		
No. 1 Brass.	$ 8.00	3/8	1/2	350 Gals	$ .50
" 2	10.00	1/2	3/4	500 "	.75
" 3	15.00	3/4	1	960 "	1.00
" 4	20.00	1	1 1/4	1,300 "	1.25
" 5	25.00	1 1/4	1 1/2	2,060 "	1.50
" 6 Iron.	35.00	1 1/4	2	3,500 "	1.75
" 7	40.00	1 1/2	2 1/2	5,000 "	......
" 8	50.00	2	3	8,000 "	......
" 9	65.00	2 1/2	4	12,000 "	......

Size No. 6 has Iron Body, balance Brass. Sizes No. 7 and 8 have Iron Bodies and Delivery connection, balance Brass. Size No. 9 has Brass Tubes, balance Iron. Sizes Nos. 6, 7, 8 and 9 made entirely of Brass to order.

Special Circulars referring to these goods, giving full information, will be sent by mail.

ENGINEERS' SUPPLIES.

Clark's Damper Regulator.

No. 1,	For 5 H. P. Boilers, ...each,	$10.00
No. 2,	" 10 to 25 H. P. Boilers, "	15.00
No. 3,	" 30 and larger, " " "	25.00

Low Pressure Damper Regulators.

Price, each...$4.50
Rubber Diaphragms for same..75

Steel Wire Flue Brush.

SIZE..	1½	2	2½	2¾	3 in.
Price, each...	1.20	1.25	1.50	1.60	1.75

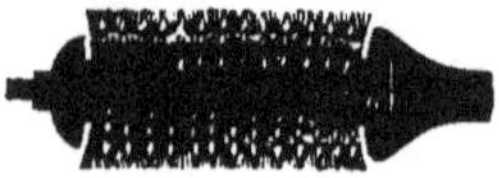

Wind's Expansion Flue Brush.

SIZE..	1½	2	2½	2¾	3	4 in.
Price, each....................................	1.50	2.00	2.50	2.75	3.00	4.00

Engineers' Favorite Flue Brush.

SIZE..	2	2½	2¾	3 in.
Price, each...	2.00	2.50	2.75	3.00

Kelsey's Tube Cleaner.

SIZE..	2	2½	2¾	3 in.
Price, each...	2.00	2.50	2.75	3.00

Ingalls' Self-Adjusting Tube Scraper and Cleaner.

SIZE	2	2½	2¾	3 in.
Price each	2.00	2.50	2.75	3.00

Steam Flue Cleaner.

No. 1, for 1 in. to 1½ in. Flues, $7.00 No. 3, for 2¾ in. to 3¼ in. Flues $9.00
No. 2, " 1¾ " 2½ " " 8.00 No. 4, " 3½ " 4 " " 10.00

Brass Trip Gongs.

SIZE	4	5	6	8	10	12	16 in.
Price, each	1.50	1.75	2.75	5.00	7.50	15.00	27.00

Gong Cranks.

Small each $ .20 Large, each $ .30

Gong Pulls.

No. 1, for 4 and 5 inch Gongs...$ .40 No. 3, " 10 inch Gongs $ 1.75
No. 2, for 6 to 8 " " ... .85 No. 4, " 12 " " and larger. 2.25

Steel and Brass Oilers.

Number	12	13	14	15	16
Steel, per doz	4.50	5.50	6.50	9.25	10.50
Brass, " "	6.50	6.50	9.20	12.00	14.00

Steel and Brass Rail Road Oilers.

Capacity	1 Pt.	1 Qt.	2 Qt.
Steel, per doz	14.00	18.00	20.00
Brass, " "	18.00	21.00	24.00

Steel and Brass Engineer's Fillers.

Capacity	1 Pt.	1½ Pt.	1 Qt.	2 Qt.
Steel, per doz.	14.00	17.00	20.00	24.00
Brass, " "	....	22.00	30.00	34 00

Engineer's Brass Oiler Sets.

Number of Pieces	5	6
Brass, per set each	6.00	9.00
Nickeled, " "	8.00	11.00

Steam Whistles.

Dia. of Bell	1	1¼	1½	2	2½	3	3½	4	5	6	8	10 in.
Screwed for Pipe	¼	⅜	⅜	¾	¾	1	1	1½	1½	2	2½	3
No. 1, Without Valve	1.76	2.00	2.50	3.25	4.50	6.00	8.50	11.00	18.00	24.00	60.00	125.00
No. 2, With Pull Up Valve	3.50	3.75	4.00	4.75	6.50	8.00	11.00	14.00	22.00	30.00	....	....
No. 3, With Pull Down Valve	.....	.....	4.60	5.50	7.00	9.00	12.50	15.00	23.00	33.00	70.00	130.00

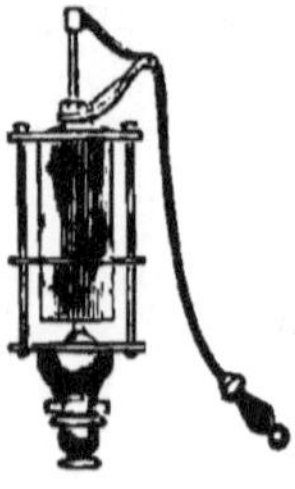

Mockingbird Steam Whistles.

Diameter of Bell	3	4	5	6 in.
Length of Bell	7	9	11	13 in.
Size of Pipe	1	1¼	1½	2 in.
Price	$40.00	$53.00	$70.00	$95.00

Chime Whistles.

Any Number or Size of Bells made to order. Prices on application.

Whistle Valves.

SIZE	3/8	1/2	3/4	1	1 1/4	1 1/2	2	2 1/2	3 in.
Price	2.00	2.25	2.75	3.25	4.00	5.50	9.50	20.00	30.00

Water Gauges. Complete.

No. 3 1/4,	Iron Wheels, Rough. Two Guards for 1/2 inch Pipe. Glass 5/8x12 in	$ 3.00
No. 3 1/2,	Iron Wheels, Finished, Two Guards for 1/2 inch Pipe, Glass 5/8x12 in	4.00
No. 7,	Wood Wheels, Finished, Two Guards for 1/2 inch Pipe, Glass 5/8x14 in	7.00
No. 10,	Wood Wheels, Finished, Four Guards for 3/4 inch Pipe, Glass 3/4x14 in	15.00

Combination Water Columns.

No. 1, Hexagon Column only, (tapped for 3/8 in. Gauge Cocks)	$2.50
No. 2, Hexagon Column only, (tapped for 1/2 in. Gauge Cocks)	4.00

Genuine Scotch Glass Tubes.

LENGTH	External Diameter.				
	1/2	5/8	3/4	7/8	1
10 inches per dozen	4.80	4.80	6.60	8.40	10.80
11 " "	4.80	4.80	6.60	8.40	10.80
12 " "	5.40	5.40	6.60	8.40	10.80
13 " "	5.40	5.40	6.60	8.40	10.80
14 " "	6.00	6.00	7.20	8.40	10.80
15 " "	6.60	6.60	7.20	9.00	10.80
16 " "	7.20	7.20	7.80	9.60	10.80
17 " "	7.80	7.80	8.40	10.20	11.40
18 " "	8.40	8.40	9.00	10.80	12.00
19 " "	9.00	9.00	9.60	11.40	13.20
20 " "	9.60	9.60	10.20	12.00	15.00
22 " "	10.80	10.80	11.40	15.00	18.00
24 " "	12.00	12.00	12.60	18.00	24.00

Utica Steam Gauges—Iron Case.

No. [illegible] 3½ inch for Air	....	$ 4.50
No. [illegible] 3½ " " Steam	....	4.50
No. 1, 5 " " "	....	6.00
No. 2, 6 " " "	....	7.50
No. 3, 6¾ " " " Pressure or Vacuum	...	9.00

BRASS CASE

No. [illegible], 3½ inch Pressure	....	$ 7.00
No. [illegible] 5 " "	....	10.75
No. 2 [illegible] " Stationary	....	13.00
No. [illegible] " High Pressure	....	14.00
No. [illegible] " Locomotive, Stationary or Vacuum	....	16.00
No. [illegible] " Pressure or Vacuum	....	23.50
No. [illegible] " " "	....	26.75
No. [illegible] " " "	....	35.00
No. 10, [illegible]		

Siphons For Steam Gauges.

Made of ¼ inch Pipe .... 25 cents each.

Nason's Steam Glue Heaters.

Numbers	1	2	3
Size of Cover. Inches.	11 x 15½	16 x 22½	16 x 28¾
Depth "	7	9	9
Size and number of Pots which can be fitted to each size,...	two 5 in. or, one 8 in. or, one 9 in.	six 5 in. or, one 10 in. and two 5 in.	two 12 in. or, one 12 in. and four 5 in. or, eight 5 in.
Price, without Pots	$8.00	$10.00	$20.00

Copper Pots for Glue Heaters.

Diameter	5	8	9	10	12 in.
Depth	5	7½	7½	8	8 in.
Price	2.00	3.50	4.00	4.50	5.50

Contact Glue Heaters.

Cabinet Bench Heater, Price	$ 3.00
No. 1 Bindery " "	3.00
No. 2 " " "	3.25

Glue Makers.

5 Gallon, Copper Body	$15.00
10 " " "	25.00

Copper Glue Pots.

Double Seamed Cabinet	$1.00
" " Gallery, No. 1 Bindery	3.00
" " " No. 2 "	3.50

Judson's Governors.

Diameter of Opening	½	¾	1	1¼	1½	2	2½	3 in.
Plain, with Speeder	16.00	18.00	20.00	22.00	25.00	30.00	40.00	50.00

Fidelity Steam Trap.

Price List, Size and Condensing Capacities, based on a pressure of 80 pounds.

No.	Lineal Feet 1 Inch Pipe	Square Feet Heat Surface	PRICE.
00	1,000	360	$22.00
0	2,000	700	26.00
1	4,000	1,400	30.00
2	7,000	2,500	40.00
3	10,000	3,500	55.00
4	15,000	5,700	75.00

Chapman's Self-Regulating Steam Trap.

Circular on application.

No. 1, Will Drain 1,500 feet 1 inch pipe	$25.00
No. 2, " 3,000 " 1 "	35.00
No. 3, " 7,000 " 1 "	60.00
No. 4, " 10,000 " 1 "	70.00

Hawes' Steam Trap.

Circular furnished on application.

Number	0	1	2	3	4	5
Tapped for Pipe	¼	½	¾	1	1¼	2 inches.
Capacity, 1 inch Pipe	100	500	1,000	2,000	4,000	20,000 feet.
Price	$6.00	$11.00	$16.00	$21.00	$26.00	$50.00

Albany Gravitating or Bucket Steam Trap.

FOR RETURNING THE WATER OF CONDENSATION DIRECTLY TO THE STEAM BOILER.

Circular forwarded on application.

No. 1, Capable of Draining 8 to 10,000 feet 1 inch Pipe	$150.00
No. 2, " " 4 to 5,000 " 1 " "	100.00
No. 3, " " 1 to 5,000 " 1 " "	75.00
Drip Tank for either size	10.00
Extra Diaphragms No. 1, 2, 3,	1.50

The Ross Pressure Regulator.

SIZE	¾	1	1¼	1½	2	3 in.
For Hot Water	7.00	10.00	14.00	20.00	24 00	60.00
For Steam	....	15.00	18.00	23.00	28.00	60.00

Crosby Water Relief Valve.

SIZE	¾	1	1¼	1½	2	2½	3	3½	4	5	6 in.
Brass	10.00	12.00	15.00	20.00	30.00	50.00	65.00	80.00	100.00	150.00	....
Iron	....	....	....	....	25.00	35.00	50.00	60.00	75.00	100.00	150.00

In ordering state pressure to be carried. If flange is desired state diameter of flange.

Crosby's Pop Safety Valves.

Prices.

¾ in. for Boilers below 5		H. P.	$10.00
1 " " from	5 to 10	"	12.00
1¼ " " "	10 to 20	"	15.00
1½ " " "	20 to 30	"	20.00
2 " " "	30 to 40	"	30.00

In ordering, state pressure to be carried.

Directions.

Setting.—Screw up or down the head-bolt which compresses the spring, until the valve opens at the pressure desired, as indicated by the steam gauge; secure the head bolt in this position by means of the lock-nut.

Caution.—Care should be taken that no red-lead, chips, or any hard substance be left in the pipes or couplings when connecting the valve with the boiler. Never make a direct connection by screwing a taper thread into the valve, but make the joint with the valve by the shoulder.

Grinding.—This valve, having flat seats on the same plane, is very easily ground by following these directions, viz: Do not grind the valve to its seats on the shell by grinding them together, but grind each part separately,—*i. e.*, grind the valve proper on a perfectly flat surface of iron or steel, until its two bearings are exactly on a plane and with good smooth surfaces; then take the shell and grind its seats in precisely the same manner; rinse both parts in water and put together, and the valve will be found to be tight; to ascertain when the bearings are on the same plane, use a good steel straight-edge.

We also furnish other makes when desired.

Watson's Steam Pressure Regulators.

FOR REDUCING AND REGULATING PRESSURE ON STEAM PIPES, &c.

Circular on Application.

	Brass Bodies, Screwed.						Iron Body.		Screwed and Flanged Iron Pipe.		
SIZES..........	1	1¼	1½	2	2½	3	2	2½	3	4	6
Price..........	17.00	22.00	28.00	38 00	55.00	70.00	38 00	55.00	70.00	90.00	150.00

McDaniel's Patent Condenser Head.

For Exhaust Steam Pipes.

SIZE........	2	3	4	5	6	8	10
Price...............	$25.00	$30.00	$40.00	$50.00	$50.00	$85.00	$120.00

PLUMBERS' BRASS WORK.

Plain Lever Bibbs.

SIZE	1/4	3/8	1/2	5/8	3/4	1	1 1/4	1 1/2	2 in.
Finished, per dozen	10.00	12.00	15.00	18.00	24.00	36.00	60.00	84.00	170.00
Nickel Plated, "	12.00	14.00	17.50	20.50	26.50	39.00	....	....	....
Rough, "	9.00	11.00	14.00	16.00	21.00	32.00	52.00	72.00	150.00

Plain Lever Bibbs.

For Iron Pipe, with Shoulder.

SIZE	1/4	3/8	1/2	5/8	3/4	1	1 1/4	1 1/2	2 in.
Finished, per dozen	11.00	13.00	16.00	19.00	26.00	39.00	64.00	90.00	180.00
Nickel Plated, "	13.00	15.00	18.50	21.50	28.50	42.00	....	....	....
Rough, "	10.00	12.00	15.00	17.00	23.00	35.00	56.00	78.00	160.00

Lever Hose Bibbs.

SIZE	1/2	5/8	3/4	1	1 1/4	1 1/2	2 in.
Finished, per dozen	16.00	19.00	26.00	39.00	64.00	90.00	180.00
Nickel Plated, "	18.50	21.50	28.50	42.00	....	....	....
Rough, "	15.00	17.00	23.00	35.00	56.00	78.00	160.00

Lever Hose Bibbs.

For Iron Pipe, with Shoulder.

SIZE	3/8	1/2	5/8	3/4	1	1 1/4	1 1/2	2 in.
Finished, per dozen	....	17.00	20.00	28.00	42.00	68.00	96.00	190.00
Nickel Plated, as above	....	19.50	22.50	30.50	45.00	....	....	....
Rough, as above	....	16.00	18.00	25.00	38.00	60.00	84.00	170.00

Plain Lever Bibb.

Screwed For Wood.

Size	¼	⅜	½	⅝	¾	1	1¼	1½	2 in.
Finished, per dozen	11.00	13.50	17.00	20.00	27.00	41 00	68.00	94.00	195.00
Rough. "	10.00	12.50	16.00	18.00	24.00	37.00	60.00	82.00	175.00

Plain Lever Bibbs.

Flange and Thimble.

Size	⅜	½	⅝	¾	1	1¼ in.
Finished, per dozen	19.00	24.00	28.00	40.00	53 00	....
Nickel Plated, "	22.00	27.50	31.50	44.00	57.00	....

Lever Hose Bibbs.

Flange and Thimble.

Size	⅜	½	⅝	¾	1	1¼ in.
Finished, per dozen	....	25 00	29.00	42.00	56.00	....
Nickel Plated, "	....	28.50	32.50	46.00	60.00	....

Plain Lever Bibb.

Flange Nut, and Bent Coupling.

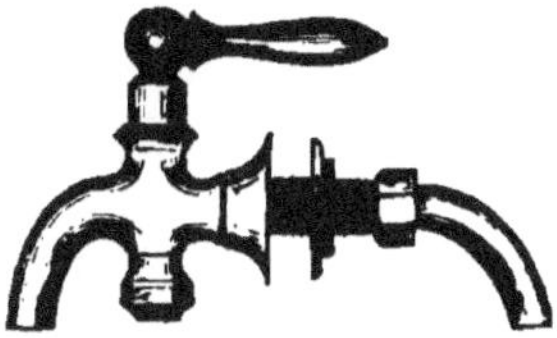

Size	½	⅝	¾ in.
Finished, per dozen	29.50	38.00	50.00
Nickel Plated, "	33.00	41.50	54.00

Lever Hose Bibbs.

Flange Nut, and Bent Coupling.

SIZE	½	⅝	¾
Finished, per dozen	30.50	39.00	52.00
Nickel Plated, "	34.00	42.50	56.00

Lever Wash Tray Bibbs.

SIZE	⅜	½	⅝	¾	1 in.
Finished, per dozen	13.00	16.00	20.00	26.00	38.00
Nickel Plated, "	15.00	18.50	22.50	28.50	41.00

Lever Wash Tray Bibbs.

For Iron Pipes, with Shoulder.

SIZE	⅜	½	⅝	¾	1
Finished per dozen	14.00	17.00	21.00	28.00	41.00
Nickel Plated, "	16.00	19.50	23.50	30.50	44.00

Lever Wash Tray Bibbs.

Flange and Thimble

SIZE	⅜	½	⅝	¾	1 in.
Finished, per dozen	20.00	25.00	30.00	42.00	55.00
Nickel Plated, "	23.00	28.50	33.50	46.00	59.00

Lever Tray Bibbs.

Flange, Nut, and Bent Coupling.

SIZE	½	⅝	¾ in.
Finished, per dozen	30.50	40.00	52.00
Nickel Plated, "	34.00	43.50	56.00

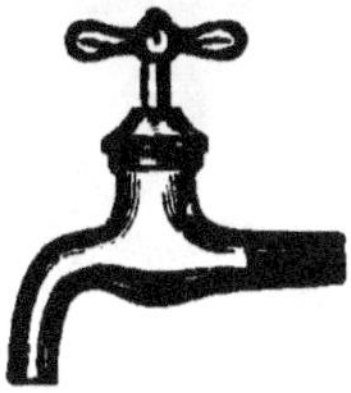

Compression Plain Bibbs.

SIZE..................	3/8	1/2	5/8	3/4	1	1 1/4	1 1/2	2 in.
Finished. per dozen.	9.00	10.00	12.00	18.00	34.00	52.00	80.00	160.00
Nickel Plated "	11.00	12.50	14.50	20.50	37.00	....	....	....
Rough. "	8.50	9.50	11.00	17.00	30.00	44.00	68.00	140.00

Compression Plain Bibbs.

For Iron Pipe With Shoulder.

SIZE..................	3/8	1/2	5/8	3/4	1	1 1/4	1 1/2	2 in.
Finished, per dozen..	10.00	11.00	13.00	20.00	37.00	56.00	86.00	170.00
Nickel Plated, "	12.00	13.50	15.50	22.50	40.00	....	....	....
Rough, "	9.50	10.50	12.00	19.00	33.00	48.00	74.00	150.00

Compression Hose Bibbs.

SIZE..................	3/8	1/2	5/8	3/4	1	1 1/4	1 1/2	2 in.
Finished, per dozen	10.00	11.00	13.00	20.00	37.00	56.00	86.00	170 00
Nickel Plated, "	12.00	13.50	15.50	22.50	40.00	....	....	....
Rough, "	9.50	10.50	12.00	19.00	33.00	48.00	74 00	150.00

Compression Hose Bibbs.

For Iron Pipe With Shoulder.

SIZE	3/8	1/2	5/8	3/4	1	1¼	1½	2 in.
Finished, per dozen.	11.00	12.00	14.00	22.00	40.00	60.00	92.00	180.00
Nickel Plated, "	13.00	14.50	16.50	24.50	43.00	....	....	....
Rough, "	10.50	11.50	13.00	21.00	36.00	52.00	80.00	160 00

Compression Plain Bibbs.

Screwed for Wood.

SIZE	3/8	1/2	5/8	3/4	1 in.
Finished	10.50	12.00	14.00	21.00	39.00
Nickel Plated	12 50	14.50	16.50	23.50	42.00
Rough	10.00	11.50	13 00	20.00	35.00

Compression Plain Bibbs.

Flange and Thimble.

SIZE	3/8	1/2	5/8	3/4	1	1¼ in
Finished, per dozen	16.00	18.00	21.00	28.00	51.00	....
Nickel Plated "	19.00	21.50	24.50	32.00	55.00	....

Compression Hose Bibbs.

Flange and Thimble.

SIZE	3/8	1/2	5/8	3/4	1	1¼ in.
Finished, per dozen	17.00	19.00	22.00	30.00	54.00	....
Nickel Plated, "	20.00	22.50	25.50	34.00	58.00	....

Compression Bibbs.

Flange, Nut, and Coupling.

SIZE	1/2	5/8	3/4 in.
Finished per dozen	25.00	32.00	44.00
Nickel Plated, "	28.50	35.50	48.00

Compression Hose Bibbs.

Flange, Nut and Coupling.

SIZE	½	⅝	¾ in.
Finished, per dozen	26.00	33.00	46.00
Nickel Plated, "	29.50	36.50	50.00

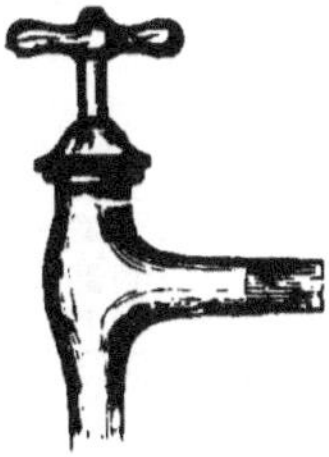

Compression Bath Bibb.

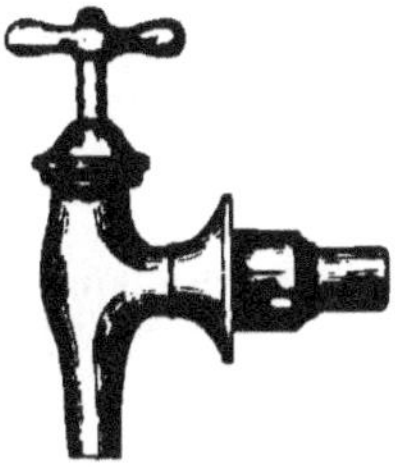

SIZE	½	⅝	¾	1 in.
For Lead Pipe, Finished per dozen,	14.00	17.00	26.00	42.00
" " Nickel Plated "	16.50	19.50	28.50	45.00
For Iron Pipe, Finished "	15.00	18.00	28.00	45.00
" " Nickel Plated "	17.50	20.50	30.50	48.00
Flange and Thimble, Finished "	22.00	26.00	36.00	59.00
" " " Nickel Plated "	25.50	29.50	40.00	63.00
Flange, Nut and Bent Coupling, Finished "	29.00	37.00	52.00	....
" " " " Nickel Plated "	32.50	40.50	57.00	....

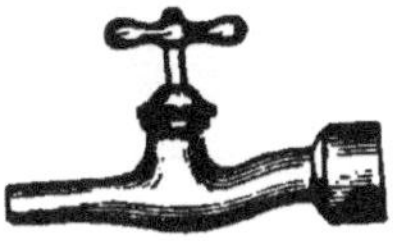

Compression Urinal Cocks.

SIZE, ½ inch.	Brass.	Nickel Plated.
Plain per dozen	18.00	21.00
Flange and Thimble "	27.00	32.00
New Pat. Jam Nut and Coupling "	36.00	40.00

Compression Urinal Cocks.

Iron Pipe.

SIZE	⅜	½	⅝ in.
Brass, per dozen	18.00	19.00	21.00
Nickeled, "	21.00	22.00	24.00

Compression Tray Bibbs.

Side Handle.

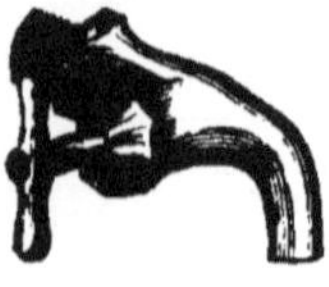

SIZE	3/8	1/2	5/8	3/4	1 in.
For Lead Pipe, Finished ......per dozen,	10.00	11.00	13.00	19 00	36.00
" " Nickel Plated ...... "	12.00	13.50	15.50	21.50	39.00
For Iron Pipe, Finished ...... "	11.00	12.00	14.00	21.00	39.00
" Nickel Plated ...... "	13.00	14.50	16.50	23.50	42.00
Flange and Thimble, Finished ...... "	17.00	19.00	22.00	30.00	53.00
" " " Nickel Plated ...... "	20.00	22.50	25 50	34.00	57.00
Flange, Nut and Bent Coup., Fin ...... "	....	26.00	33.00	46.00	....
" " " " Nickel Plated, "	....	29.50	36.50	50.00	....

Compression Tray Bibbs.

Straight.

SIZE	3/8	1/2	5/8	3/4	1 in.
For Lead Pipe, Finished ......per dozen,	10.00	11.00	13.00	19.00	36.00
" " Nickel Plated ...... "	12.00	13.50	15.50	21.50	39.00
For Iron Pipe, Finished ...... "	11.00	12.00	14.00	21.00	39.00
" " Nickel Plated ...... "	13.00	14.50	16.50	23.50	42.00
Flange and Thimble, Finished ...... "	17.00	19.00	22.00	30.00	53.00
" " " Nickel Plated "	20.00	22.50	25 50	34.00	57.00

Tray Bibbs.

Flange, Nut and Bent Coupling.

SIZE	1/2	5/8	3/4	1 in.
Finished	26.00	33.00	46.00	77.00
Nickeled	29.50	36.50	50.00	....

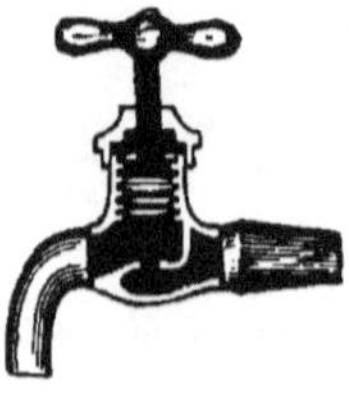

Compression Plain Bibbs.

Grundy Pattern,—Sectional View.

This illustration shows the construction of Grundy Compression Bibbs, which we can furnish in all the regular varieties of 1/2, 5/8 and 3/4 inch sizes. Except Tray Bibbs.

Compression Bibbs, Lead or Iron Pipe With Stuffing Box.

For Stuffing Box Add to Regular List.

Sizes	1/2	5/8	3/4	1 in.
Per dozen	2.00	2.00	3.00	4.00

Compression Double Bath Cock.

Fig. 35—With Sprinkler and Hose Coupling.

Finished, each $5.00
Nickel Plated, each 5.50

Compression Double Bath Cock.

Fig. 36—With Sprinkler and Hose Coupling—Same as Fig. 35.

Finished, each .... $7.00 Nickel Plated, each $7.50

Compression Double Bath Cock.

Fig. 37.—Handles on Top.

WITH SPRINKLER AND HOSE COUPLING.

Finished, each $7.00
Nickel Plated, each 7.50

Compression Double Bath Cock.

Fig. 39.

Finished, each .... $7.50 Nickel Plated, each $8.00

Swing Basin Cocks.

No. 0.

Finished,	per dozen	$18.00
Nickel Plated,	"	21.00

No. 1.

Finished,	per dozen	$18.00
Nickel Plated,	"	21.00

No. 2.

Finished,	per dozen	$22.00
Nickel Plated,	"	25.00

No. 4.

Finished,	per dozen	$24.00
Nickel Plated,	"	27.00

Compression Basin Cocks.

No. 0.

Finished,	per dozen	$16.00
Nickel Plated,	"	19.00
Add for 4 Arm Handle,	per dozen	1.00
" Large Tube,	"	2.00
" Stuffing Box,	"	2.00

No. 1.

Finished, per dozen	$17.00
Nickel Plated, "	20.00
Add for 4 Arm Handle, per dozen	2.00
" Large Tube, "	2.00
" Stuffing Box, "	2.00

No. 2.

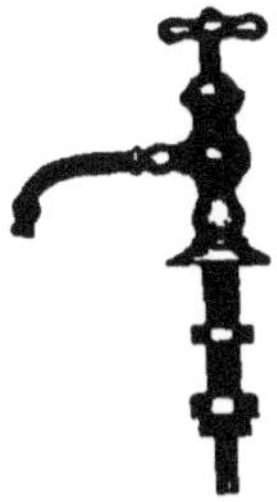

Finished, per dozen	$22.00
Nickel Plated, "	25.00
Add for 4 Arm Handle, Plain, per dozen	2.00
" " " Fancy, "	4.00
" Large Tubes, per dozen	2.00
" Stuffing Box, "	2.00

No. 3.

Finished, per dozen	$28.00
Nickel Plated, "	32.00
Add for Large Tubes, per dozen	2.00
" Stuffing Box, "	2.00
Less with T Handle, "	4.00

No. 4.

Finished per dozen	$20.00
Nickel Plated. "	24.00
Add for 4 Arm Handle, per dozen	4.00
" Large Tubes, "	2.00
" Stuffing Box. "	2.00

No. 4½.

Finished, per dozen	$26.00
Nickeled, "	30.00
Add for Stuffing Box	2.00

No. 7.—GRUNDY PATTERN.

Same as No 7½, but without Stuffing Box.

Finished, per dozen	$24.00
Nickel Plated, "	28.00

No. 7½.

Grundy Pattern, with Stuffing Box.

Finished, per dozen	$26.00
Nickel Plated, "	30.00
Add for 4 Arm Handle, per dozen	4.00

Double Compression Basin Cocks.

With Ring Cup.

Finished, each	$ 9.00
Nickel Plated, each	10.00

Bracket Swing Basin Cocks.

Plain, Finished,	per dozen	$18.00
" Nickel Plated,	"	22.00
Octagon, Finished,	"	34.00
" Nickel Plated,	"	40.00

Compression Double Bracket Basin Cocks.

Or With Four Arm Handle.

Finished, each	$5.00
Nickel Plated, each	5.50

Pantry Cocks.

Lever Handle, Ground Key.

		Brass	Nickel Plated.
No. 1, Small, Plain,	per dozen	30.00	34.00
" Small, Hose End,	"	33.00	37.00
No. 2, Large, Plain,	"	34.00	38.00
" Large, Hose End,	"	37.00	41.00

Compression Pantry Cocks.

		Brass.	Nickel Plated.
No. 1, T Handle, Small, Plain,	per dozen.....	30.00	34.00
" " " Hose End,	"	33.00	37.00
No. 2, " " Large, Plain,	"	34.00	38.00
" " " Hose End,	"	37.00	41.00
Add for 4 Arm Handle, both Sizes,	"	2.00	2.00

Compression Double Pantry Cocks.

Finished, each $11.00
Nickel Plated, each 12.00

Double Compression Shampooing and Basin Cocks.

Finished, each $15.00
Nickel Plated 17.00

Shampooing Sprinklers.

Finished, per dozen .... $ 8.00
Nickel Plated, " 10.00

Chain Stays.

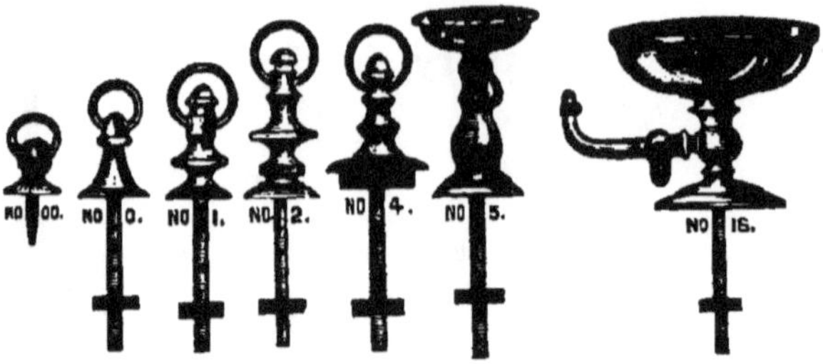

NUMBER	00	0	1	2	4	5	16
Brass, per dozen	2.00	2.00	3.00	3.50	3.50	8.00	24.00
Nickel Plated "	2.50	2.50	3.75	4.25	4 25	9.00	27.00

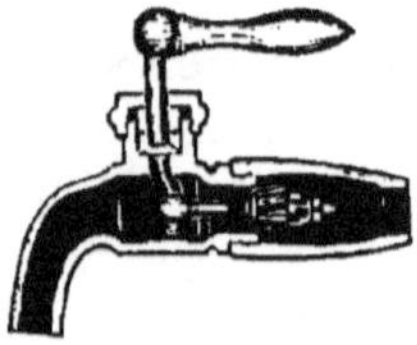

Sectional View of Improved Bibbs.

Improved Bibbs.

Lead Pipe.

SIZE	½	⅝	¾	1 in.
Plain, Finished, per dozen	18.00	20.00	26.00	36.00
" Nickeled, "	22.00	24.00	32.00	46.00
Hose, Finished, "	21.00	24.00	30.00	38.00
" Nickeled, "	25.00	28.00	36.00	48.00

Screwed for Iron Pipe.

SIZE	½	⅝	¾	1 in.
Plain, Finished, per dozen	21.00	24.00	30.00	40.00
" Nickeled, "	25.00	28.00	35.00	48.00
Hose, Finished, "	24.00	27.00	32.00	42.00
" Nickeled, "	28.00	31.00	37.00	50.00

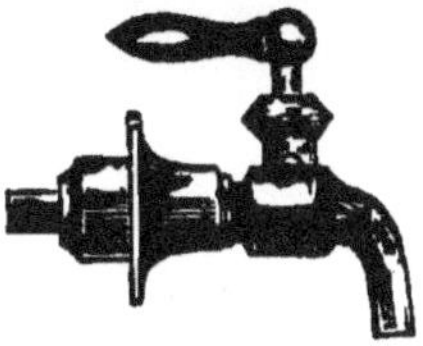

Improved Flange and Thimble Bibbs.

SIZE	½	⅝	¾	1 in.
Plain, Finished	26.00	28.00	36.00	48.00
" Nickeled	32.00	34.00	42.00	58.00
Hose, Finished	29.00	32.00	40.00	50.00
" Nickled	35.00	38.00	46.00	60.00

Improved Bath Bibbs.

SIZE		½	⅝	¾	1 in.
Plain, Finished,	per doz.	18.00	20.00	26.00	36.00
" Nickeled,	"	22.00	25.00	32.00	42.00
Flanged and Thimble, Finished,	"	25.00	28.00	34.00	48 00
" " " Nickeled,	"	31.00	34.00	40.00	56.00
Flanged, Nut and Coupling, Finished,	"	34.00	40.00	52.00	72.00
" " " Nickeled,	"	40.00	46.00	58.00	80.00

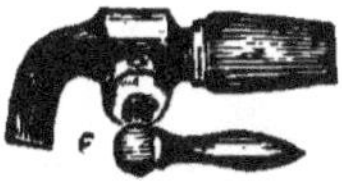

Improved Wash Tray Bibbs.

SIZE		½	⅝	¾	1 in.
Plain, Finished,	per doz	20.00	22.50	30.00	40.00
" Nickeled,	"	24.00	26.50	36.00	50.00
Flanged and Thimble, Finished,	"	26.00	28.00	36.00	50.00
" " " Nickeled,	"	32.00	34.00	42.00	60.00
Flgd., Nut and Bent Coup., Finished,	"	35.00	40.00	54.00	74.00
" " " Nickeled,	"	41.00	46.00	60.00	84.00

Improved Stop With Coupling.

SIZE	½	⅝	¾	1	1¼ in.
Rough, per dozen	18.00	21.00	25.00	36.00	60.00
Finished, "	20.00	24.00	28.00	40.00	65.00
Nickeled, "	25.00	30.00	34 00	50 00	......

Improved Basin Cocks.

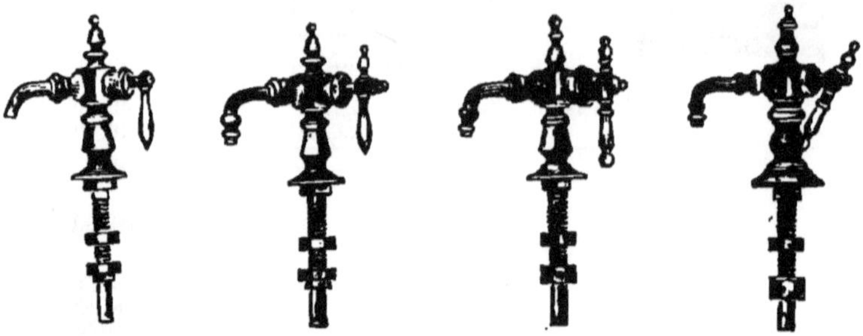

No. 1. No. 2. No. 3. No. 4.

NUMBER	1	2	3	4
Finished per dozen	33.00	36.00	46 00	56.00
Nickeled "	40.00	44.00	54.00	64.00

Double Basin Cock.

No. 8.

Finished, each	$12 00
Nickeled, "	13.00

Improved Pantry Cock.

No. 1.

Plain, Finished	per dozen	$36.00
" Nickeled	"	42.00
Hose End, Finished,	"	38 00
" " Nickeled,	"	45.00

Double Pantry Cock

No. 9.

Finished, each	$12.00
Nickeled, "	13.00

Improved Double Bath Cock.

No. 4½.

Finished	$10.50
Nickeled	12.00

Basin Waste Cock.

Finished, each	$6.00
Nickeled, "	7.00

Bath Waste Cock.

Finished, each	$8.00
Nickeled, "	9.00

Trimmings for Improved Bibbs.

SIZE	½	⅝	¾	1 in.
Eccentrics, per dozen	3.00	3.00	4.00	6.00
Valve and Stem, "	3.00	3.00	4.00	6.00
Rubber Valves, "	.75	.75	1.00	1.25
Handles, "	3.00	3.00	4.00	6.00

Moore' Patent Self-Closing Cocks.

SIZE	½	⅝	¾ in.
Plain Bibbs, for Lead Pipe, per dozen	24.00	27.00	33.00
" " " Iron " "	28.00	31.00	37.00
Hose " " Lead " "	27.00	30.00	36.00
" " " Iron " "	31.00	34.00	40.00
Flange and Thimble, "	36.00	39.00	....
" " " Hose End, "	40.00	43.00	....
Flange, Nut and Bent Coupling, "	45.00	48.00	....
Stops for Lead Pipe, "	24.00	27.00	33.00
" " Iron " "	32.00	35.00	35 00
" " " " One End, "	28.00	....	31.00
Add for Nickel Plating, "	4.00	4.00	5.00
Hopper Cocks, for Lead Pipe, "	27.00	30.00	....
" " " Iron " "	31.00	....	34.00
Add for Nickel Plating, "	4.00	4.00	5.00

	Finished.	Nickel Plated.
Basin Cocks per dozen	45.00	48 00
Pantry " "	54.00	64.00
" " Hose End, "	58.00	68.00
Urinal, " "	36.00	41.00
" " Iron Pipe. "	40.00	45.00

Springs for Moore's Self Closing Bibbs $6.00 per dozen.

Telegraph Bibbs.

SIZE	⅜	½	⅝	¾ in.
Finished, for Lead Pipe, per doz.	15.00	17.00	20.00	26.00
" " Iron " "	16.00	18 00	21.00	28.00
Nickel Plated, for Lead Pipe, "	17.00	19.50	22.50	28.50
" " " Iron " "	18.00	20.50	23.50	30.50

Telegraph Basin Cocks.

	Brass.	Nickel Plated.
Per dozen	24.00	27.00
" " add for Large Tubes	2.00	2.00

Plain Finished Stops.

Size	¼	⅜	½	⅝	¾	1	1¼	1½	2 in.
Finished, per doz.	10.50	12.50	15.50	18.50	25.00	37.00	62.00	86.00	175.00
Nickel Plated, "	12.50	14.50	18.00	21.00	27.50	40.00	....	....	....

Plain Finished Stops.

Screwed or Iron Pipe.

Size	¼	⅜	½	⅝	¾	1	1¼	1½	2 in.
Finished, per doz.	11.50	13.50	16.50	20.50	27.00	40.00	58.50	90.00	185.00
Nickel Plated, "	13.30	15.50	19.00	23.00	29.50	43.00	....	....	....

Plain Finished Stops.

With Waste.

Size	⅜	½	⅝	¾	1 in.
Finished, per dozen	14.00	17.00	20.00	27.00	40.00
Nickel Plated, "	16.00	19.50	22.50	29.50	43.00

Plain Finished Stops—With Waste.

Screwed for Iron Pipe.

SIZE	3/8	1/2	5/8	3/4	1 in.
Finished, per dozen	15.00	18.00	22.00	29.00	43.00
Nickel Plated, "	17.00	20.50	24.50	31.50	46.00

Lever Handle Shower Stops.

Flange and Handle.

SIZE	1/2	5/8	3/4 in.
For Lead Pipe, Finished, Flange and Handle, per dozen	24.00	29.00	36.00
" " " Nickel Plated, " " "	26.50	31.50	38.50
For Iron Pipe, Finished, Flange and Handle, per dozen	25.00	31.00	38.00
" " " Nickel Plated, " " "	27.50	33.50	40.50

Rough Stops.

Lever or T Handle.

SIZE	1/4	3/8	1/2	5/8	3/4	1	1 1/4	1 1/2	1 3/4	2 in.
Per dozen	7.00	9.00	12.00	15.00	19.00	28.00	46.00	64.00	85.00	110.00

Rough Stops.

Lever or T Handle, Tube Waste, Stop on Key.

SIZE	1/4	3/8	1/2	5/8	3/4	1	1 1/4	1 1/2	2 in.
Per dozen	8.00	10.00	13.00	16.00	20.50	30.00	49.00	68.00	120.00

Rough Stops.

Lever or T Handle, Screwed for Iron Pipe.

SIZE	1/4	3/8	1/2	5/8	3/4	1	1 1/4	1 1/2	2 in.
Per dozen	8.00	10.00	13.00	17.00	21.00	31.00	50.00	70.00	120.00

Rough Stops.

Lever or T Handle, Screwed for Iron Pipe, Tube Waste, Stop on Key.

SIZE	1/4	3/8	1/2	5/8	3/4	1	1 1/4	1 1/2	2 in.
Per dozen	9.00	11.00	14.00	18.00	22.50	33.00	53.00	74.00	130.00

Rough Stops.

Lever or T Handle, for Lead and Iron Pipe.

SIZE	1/4	3/8	1/2	5/8	3/4	1	1 1/4	1 1/2	2 in.
No Waste, per dozen	7.50	9.50	12.50	16.00	20.00	29.50	48.00	67.00	115.00
With " "	8.50	10.50	13.50	17.00	21.50	31.50	51.00	71.00	125.00

Rough Stops.

Round Way, Lever or T Handle.

SIZE	3/8	1/2	5/8	3/4	1	1 1/4	1 1/2	1 3/4	2 in.
Per dozen	15.00	17.00	20.00	25.00	44.00	70.00	100.00	140.00	180.00

Rough Stops.

Round Way, Lever or T Handle, Tube Waste, Stop on Key.

SIZE	3/8	1/2	5/8	3/4	1	1 1/4	1 1/2	1 3/4	2 in.
Per dozen	16.00	18.00	21.00	26.50	46.00	73.00	104.00	148.00	190.00

Rough Stops.

Round Way, Lever or T Handle, Screwed for Iron Pipe.

SIZE	3/8	1/2	5/8	3/4	1	1 1/4	1 1/2	2 in.
Per dozen	16 00	18.00	22.00	27.00	47.00	74.00	106.00	190.00

Rough Stops.

Round Way, Lever or T Handle, Screwed for Iron Pipe, Tube Waste, Stop on Key.

SIZE	3/8	1/2	5/8	3/4	1	1 1/4	1 1/2	2 in.
Per dozen	17.00	19.00	23.00	28.50	49.00	77.00	110.00	200.00

Round Way Rough Stops.

Fig. 72.

H Pattern, for Iron Pipe, T or Lever Handle.

SIZE	1/2	3/4	1	1 1/4 in.
No Waste, Light Pattern, per dozen	12.00	18.00	30.00	....
" " Heavy " "	15.00	20.00	40.00	65.00
With " Light " "	13.00	19 50	32.00	....
" " Heavy " "	16.00	21.50	42.00	68.00

Fig. 72.

H Pattern, Rough Stops, Lead Pipe.

SIZE	1/2	5/8	3/4	1 in.
T or Lever Handle, No Waste	11.00	15.00	16.00	27.00
" " " With "	12.00	16.00	17.50	29.00

Rain and Well Water Cocks.

SIZE	3/4	1	1 1/4	1 1/2	2 in.
For Lead Pipe, per dozen	50.00	70.00	84.00	120 00	200.00
" Iron " "	56 00	79.00	96.00	138.00	230.00

Compression Stops.

SIZE	3/8	1/2	5/8	3/4	1	1 1/4	1 1/2 in.
Rough, per dozen	9.50	10.50	12.00	18.50	32.00	48.00	84 00
Finished, "	10.00	11.00	13.00	19.50	36.00	56.00	96.00
Nickel Plated, "	12.00	13.50	15.50	22.00	39.00	58.00	....

Compression Stops and Waste.

With Stuffing Box. Lead Pipe.

SIZE	1/2	5/8	3/4	1 in.
Rough, per dozen	14.00	17.00	24.00	42.00
Finished, "	15.50	18.50	25.50	44.00
Nickel Plated, "	18.00	21.00	28.00	47.00

Compression Stops—For Iron Pipe.

SIZE	3/8	1/2	5/8	3/4	1	1 1/4 in.
Rough, per dozen	10.50	11.50	14 00	20.50	35.00	52.00
Finished, "	11.00	12.00	15.00	21.50	39.00	60.00
Nickel Plated, "	13.00	14.50	17.50	24 00	42.00	....

Compression Stops and Waste.

For Iron Pipe, with Stuffing Box.

SIZE	1/2	5/8	3/4	1 in.
Rough, per dozen	15.00	19.00	26.00	45.00
Finished, "	16.50	20.50	27.50	47.00
Nickel Plated, "	19.00	23.00	30.00	50.00

Compression Stops.

Finished Flange and Handle.

SIZE	3/8	1/2	5/8	3/4
For Lead Pipe Finished, per doz.	18.00	22.00	26.50	36.00
" " " Nickel Plated "	20.00	24.50	29.00	38.00
" Iron " Finished, "	19.00	23.00	28.50	38.00
" " " Nickel Plated, "	21.00	25.50	31.00	40.50

Compression Stops.

For Lead and Iron Pipe.

SIZE	3/8	1/2	5/8	3/4	1 in.
Rough, per dozen	10.00	11.00	13.00	19.50	33.50
Finished, "	10.50	11.50	14.00	20.50	37.50
Nickeled, "	12.50	14.00	16.50	23.00	40.00

Compression Stops.

Lead or Iron Pipe, with Stuffing Box.

For Stuffing Box add to Regular List.

SIZE	3/8	1/2	5/8	3/4	1 in.
Per dozen	1.50	2.00	2.00	3.00	4.00

Compression Sill Cocks.

Size 3/4 or 1/2, Finished, per dozen	$28.00
" " Nickel Plated, "	32.00

Hydrant Cocks.

Nut and Washer.

SIZE	1/2	5/8	3/4	1 in.
For Lead Pipe, per dozen	14.00	17.00	21.00	33.00
" " and Iron Pipe, "	14.50	18.00	22.00	34.50
" Iron Pipe, "	15.00	18.50	23.00	36.00

Compression Hydrant Cocks.

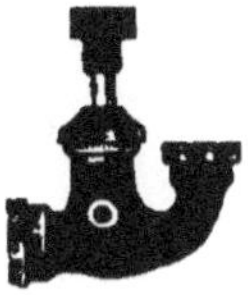

SIZE	1/2	5/8	3/4	1 in.
For Lead Pipe, per dozen	14.00	17.00	22.00	38.00
" Iron " "	15.00	19.00	24.00	41.00

Garden Hose Valves.

SIZE	1/2	3/4	1	1¼	1½	2 in.
Rough, each	1.25	1.65	2.20	3.40	4.90	7.20

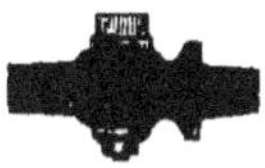

Corporation Stops

Both Ends Screwed, Male or Female, for Iron Pipe.

SIZE	3/8	1/2	5/8	3/4	1	1¼	1½ in.
Per dozen	13.00	16.00	20.00	29.00	46.00	90.00	110.00

Corporation Stops.

With Bent or Straight Couplings.

SIZE	3/8	1/2	5/8	3/4	1	1¼ in.
For Iron Pipe, Male or Female Thread,	16.00	19.00	23.00	34.00	53 00	104 00

Corporation Stops.

SIZE	3/8	1/2	5/8	3/4	1	1¼ in.
For Hubbell's, Sperring's, or Mueller's Tapping Machines	16.00	19.00	23.00	34.00	53.00	104.00

Payne's Eclipse Tapping Machine.

For Tapping under Pressure.

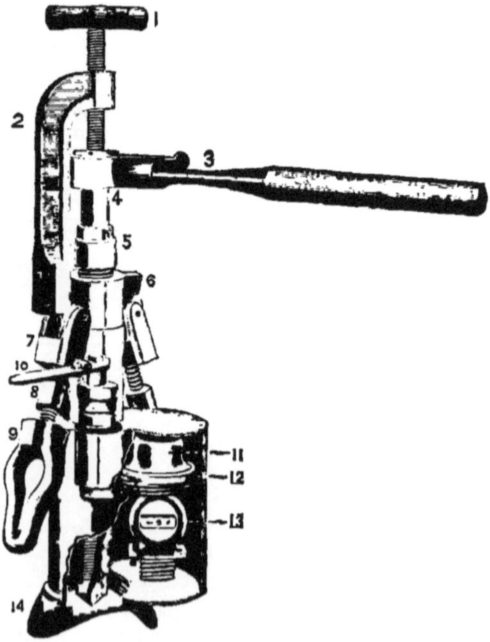

Every Machine tested to 600 Lbs. to the Sq. in. and Guaranteed.

THE MACHINE TAPS ON THE TOP OR SIDE OR AT ANY ANGLE DESIRED.

PRICE LIST.

No. 1 complete with 3/8, 1/2, 5/8, 3/4 and 1-inch taps and four saddles..........$100.00
No. 2 complete with 1, 1¼, 1½, and 2-inch taps " " " 150.00

The No. 2 Machine will tap all the small sizes by getting the taps and mandrels extra.

PRICE LIST OF EXTRA TOOLS.

3/8-inch tap, $3.00; ½-inch tap, $4.00: 5/8 and 3/4-inch taps, 4.50; 1-inch tap $5.00; 1¼-inch tap, $6.00; 1½-inch tap, $7.00; 2-inch tap, $8.00; Mandrels, $2.00; saddles, $1.00; chain, $1.00: tightening bolts, 50 cents each: harps 25 cents, each; clevises, 25 cents, each; feeder yoke, $1.00; steel feeder screws, $1.00; malleable wrenches, 25 cents; gaskets 10 cents each.

Payne's Improved "Daisy" Dry Pipe Tapping Machine.

FOR LOW PRESSURE GAS MAINS.

No Escape of Gas while Drilling and Tapping.

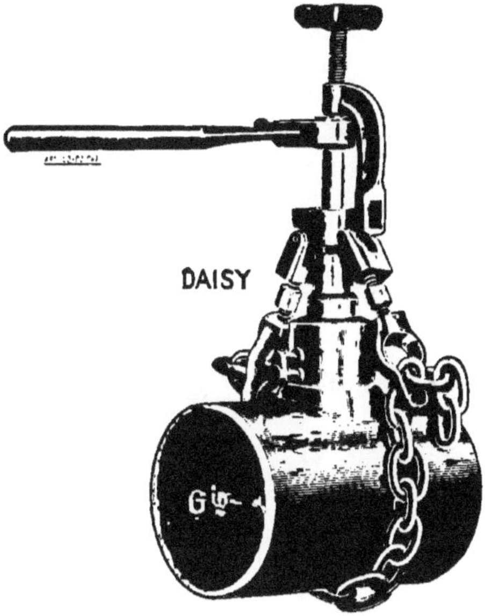

PRICE LIST.

Machine, with one Saddle, Chain, Feeder Yoke and Ratchet..................$20.00

By ordering chain proper length, and extra saddles, this machine will tap ANY SIZE OF PIPE.

Extra Saddles, each	$1.00
¼ inch Taps and Drills	3.00
3/8 inch Taps and Drills	3.00
½ inch Taps and Drills	4.00
¾ inch Taps and Drills	4.50
1 inch Taps and Drills	5.00
1¼ inch Taps and Drills	6.00
1½ inch Taps and Drills	7.00
2 inch Taps and Drills	8.00

Payne's Corporation Stops.

Female, for Iron Pipe.

SIZE	3/8	1/2	5/8	3/4	1	1 1/4	1 1/2	2 in.
Price, per dozen	12.00	13.20	16.80	25.20	40.20	80 00	100.00	180.00

Male, for Iron Pipe.

SIZE	3/8	1/2	5/8	3/4	1	1 1/4	1 1/2	2 in.
Price, per dozen	12 00	13.20	16.80	25.20	40.20	80.00	100.00	180.00

With Straight Tail Piece, for Lead Pipe.

SIZE	3/8	1/2	5/8	3/4	1	1 1/4	1 1/2	2 in.
Price, per dozen	14.40	16.20	26.40	30.00	46.20	96 00	128.00	220.00

With 1-8 Bend for Lead Pipe.

SIZE	3/8	1/2	5/8	3/4	1	1 1/4	1 1/2	2 in.
Price, per dozen	14.40	16.20	20.40	30.00	46.20	96.00	128.00	220.00

Fell or Fish Stops. 1-4 Bend, for Lead Pipe.

SIZE	3/8	1/2	5/8	3/4	1	1 1/4	1 1/2	2 in.
Price, per dozen	14.40	16.20	20.40	30.00	46.20	96 00	128.00	220.00

Ball Cocks.

Ground Key.

SIZE	3/8	1/2	5/8	3/4	1	1 1/4 in.
For Lead Pipe, per dozen	10.00	12.00	15.00	19.50	30.00	50.00
" Iron " "	11.00	13.00	16.50	21.50	33.00	54.00

Compression Ball Cocks.

SIZE	3/8	1/2	5/8	3/4	1
For Lead Pipe, per dozen	8.50	9.50	11.00	17.00	30.00
" Iron " "	9.50	10.50	12.00	19.00	33.00

Berkley Ball Cock.

Top or Bottom Supply.

Size, one half inch, for Lead Pipe $1.50

Peck's Improved Ball Cocks.

SIZE		1/2	5/8	3/4	1	1 1/4 in.
Plain,	per dozen	12.00	14.00	20.00	34.00	56.00
With Stop,	"	20.00	22.00	32.00	52.00	92.00
Copper Balls, 6 inch,	"	9.00	......	......	......	......

Compression Hopper Cocks.

Angle.

SIZE		1/2	5/8	3/4
Finished, Flange and Handle,	per dozen	16.00	19.00	24.00
Nickel Plated, Flange and Handle,	"	18.50	21.50	26.50

Compression Hopper Cocks.

Angle.

SIZE	1/2	5/8	3/4 in.
Finished, Lead and Iron, per dozen	17.00	20.00	26.00
Nickeled, " " "	19.50	22.50	28.50

Compression Hopper Cocks.

Straight.

SIZE	½	⅝	¾
Finished, Flange and Handle, per dozen	18.00	21.00	28.00
Nickel Plated, Flange and Handle, "	20.50	23.50	30.50

Compression Hopper Cocks.

Straight.

SIZE	½	⅝	¾
Finished, Lead and Iron Pipe, per dozen	18.50	22.00	29.00
Nickeled, " " " " "	21.00	24.50	31.50

Hopper Valves.

Bartholomew's Single, per dozen	$26.00

Rough Beer Cocks.

With Coupling.

SIZE	½	⅝	¾
To Drive, per dozen	16.00	21.00	32.00
" Screw, "	17.00	22.50	34.00

Rough Beer Cocks.

Key on Side.

Per dozen	$36.00

No. 2, Bar Cocks.

With Straight or Bent Coupling.

Finished,	per dozen	$59.00
Nickel Plated,	"	56 00

Counter Cocks.

	Brass.	Nickel
Round, T Handle, per dozen...........	30.00	34.00
" Lever " "	32.00	36.00
Octagon, T " "	34.00	38.00
" Lever " "	36.00	40.00

Under Counter Cocks.

Rough, per dozen..	$18.00
Finished. "	20.00

Beer Pumps.

Small, price each..	$6.00
Large, "	6.50

We also furnish to order any of the regular styles of Air Pressure Pumps and apparatus for lager and ale.

Beer Pump Stands.

Iron, price each..	$1 00

Cooler Cocks.

Fig. 136.

Size	1/4	3/8	1/2	5/8
Finished, per dozen..................................	6.00	7.00	9.00	12.00
Nickel Plated, "	7.50	8.50	10.50	14.00

Cooler Cocks.

Fig. 137.

Size..................................	1/4	3/8	1/2	5/8
Finished, per dozen.........	7.00	7.50	9.00	12.00
Nickeled Plated, "	8.50	9 00	10.50	14.00

Self-Closing Cooler Cocks.

SIZE	1/4	3/8	1/2	5/8
Finished, per dozen	11.00	12.00	14.00	17.00
Nickel Plated, "	13.00	14.00	16.50	19.50

Self-Closing Cooler Cock with Thimble.

SIZE	3/8	1/2	5/8
Finished, per dozen	14.00	16.00	19.00
Nickel Plated, "	16.00	18.50	21.50

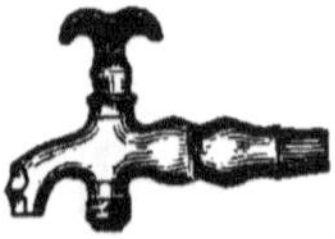

Urn Cocks.

SIZE	1/4	3/8	1/2	5/8 in.
Brass, T Handle, Finished, per dozen	10.00	12.00	17.00	....
" " " Nickel Plated, "	12.00	14.00	19.50	....
Rubber, " " Finished, "	12.00	14.00	20.00	....
" " " Nickel Plated, "	14.00	16.00	22.50	....
Ivory " " Finished "	20.00	21.00	28.00	....
" " " Nickel Plated, "	22.00	23.00	30.50	....
Brass, 4 Arm " Finished, "	13.00	15.00	19.50	23.00
" " " Nickel Plated, "	15.00	17.00	22.00	25.50

Compression Oil Cocks.

SIZE	1/4	3/8
Screw Shank, per dozen	6.00	7.00
Plain " "	5.50	6.50

Racking Cocks.

To Drive.

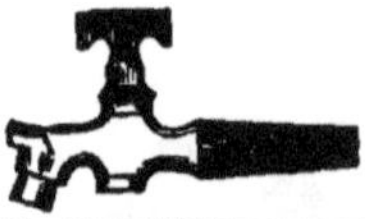

SIZE	1/4	3/8	1/2	5/8	3/4	1	1 1/4	1 1/2 in.
Finished, per dozen	5.00	7.50	10.00	14.50	17.00	30.00	54.00	72.00

Racking Cocks.

To Screw.

SIZE	1/4	3/8	1/2	5/8	3/4	1	1 1/4	1 1/2 in.
Finished, per dozen	5.50	8.50	11.00	16.00	19.00	32.50	58.00	77.00

Lock Cocks.

SIZE	3/8	1/2	5/8	3/4 in.
Finished, to drive	12.00	14.00	18.00	24.00
" " screw	12.50	15.00	19.50	26.00

Globe Cocks.

T Handle.

SIZE	1/4	3/8	1/2	5/8	3/4	1 in.
Tinned End, Finished, per dozen	8.00	10.00	13.50	16.00	22.00	33.00
To Screw, " "	8.50	10.50	14.50	17.50	24.00	35.50

Globe Cocks.

Lever Handle.

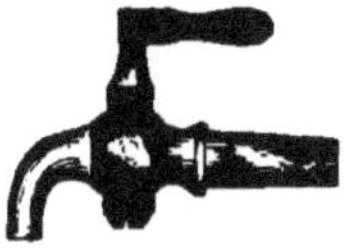

SIZE	1/4	3/8	1/2	5/8	3/4	1 in.
Tinned End, Finished, per dozen	9.00	11.00	14.50	18.00	24.00	36.00
To Screw, " "	9.50	11.50	15.50	19.50	26.00	38.50

Powell' Patent Basin, Bath and Sink Plugs.

With Rubber Stoppers.

		Brass	Nickel Plated.
Common Overflow Basin Plugs,	per dozen	7.50	9.00
Patent " " "	"	9.00	10.50
Rubber Stoppers only, 7/8, 1, or 1 1/8 inch,	"	......	1.00
" " " 1 1/4 inch,	"	.....	1.25
" " " 1 1/2 "	"	......	1.75
" " " 2 "	"	......	2.50

SIZES		1	1 1/4	1 1/2	2 in.
Bath Plugs, Brass,	per dozen	2.00	2.25	3.50	5.00
" " Nickel Plated,	"	2.50	2.75	4.00	6.00
Wash Tray Plugs, Brass,	"	2.50	3.25	4.00	6.00
" " " Nickel Plated,	"	3.50	4.25	5.00	7.50
Soap Stone Tray Plugs, Brass,	"	8.00	9.00	13.50	20.00
" " " " Nickeled,	"	9.00	10.50	15.00	22.50

Sewer Valves

SIZE	2 1/2	3	4	5	6	8 in.
Price, each	2.00	2.50	3.50	8.00	10.00	18.00

Bottle Washers.

Price, each	$5.00
" " Large Flange for Tumblers	6.50

Basin Plugs.

	Brass.	Nickel Plated.
Common Overflow, per dozen	8.00	8.50
Patent, " "	9.00	9.50

Sink Plug and Coupling.

Patent Overflow for Earthen Sinks.

SIZE..	1	1¼	1½ in.
Brass, per dozen........................	15.00	17.00	30 00
Nickel Plated, "	17.00	20.(0	33.00

Plug and Coupling.

For Soapstone Wash Trays.

SIZE..	1	1¼	1½ in.
Brass, per dozen........................	15.00	16 00	26.00
Nickeled, "	17.00	19.00	29.00

Basin Grates.

	Brass.	Nickel Plated.
Per dozen..	12.00	13 00

Bath Waste and Washer.

Size............	1	1¼	1½	2	2½	3	3½	4 in.
Per dozen.........	8.00	9.00	12 00	15.00	24.00	30.00	44.00	60.00

Bath or Sink Plugs.

SIZE....................	1	1¼	1½	2	2¼	2½	3	4 in.
Brass, per dozen,...	2.50	3.50	4.50	8.00	10.00	15.00	18.00	36.00
Nickel Plated " ...	3.00	4.00	5.00	9.00	12.00	17.00	21.00	40.00

Tray Plugs.

For Wooden Sinks.

SIZE......	3/4	1	1 1/4	1 1/2	1 3/4	2	2 1/4	2 1/2	3	4 in.
Per dozen	3.50	3.75	6.00	7.00	9.00	10.00	13 00	17.00	22.00	44.00

Trap Screws.

SIZE............	3/4	1	1 1/4	1 1/2	2	2 1/2	3	3 1/2	4 in.
Per dozen..........	2.50	3.00	3.50	4.50	8.00	12.00	18.00	26.00	30.00

Basin Clamps.

NUMBER..	2	3	4
Per dozen..	1.25	1.50	2.00

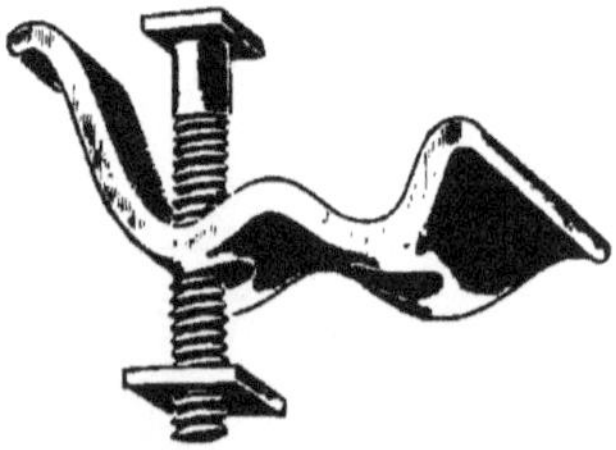

Self Adjusting.

Per dozen...$1.50

Reversible Water Filters.

No. 1, Finished, per doz. $15.00 | Nickel Plated, per doz. $17.00

Globe Reversible Filters.

No. 2, Nickel Plated, per dozen.................................$20.00

Croton Filters.

Finished, per dozen...$5.00
Nickel Plated " 6.00

Brass End Ferrules.

For Iron Pipe.

SIZE	2	3	4	5	6 in.
Per dozen	5.00	9.00	11.00	27.00	36.00
2 × 1½ and 2 × 1¼ Reducing, per doz.	5.00				

End Ferrules.

With Trap Screws.

SIZE	2	3	4 in.
Per dozen	10.00	20.00	30.00

1-8 Bend Brass Ferrules.

SIZE	2	3	4 in.
Price per dozen	12.00	18.00	24.00

Lead End Ferrules.

SIZE	4 in. Long.	6 in. Long.	8 in. Long.	10 in. Long.	12 in. Long.
1½ x 2, Regular	.23	.35	.45	.50	.55
2 "	.18	.30	.40	.45	.50
3 "	.25	.40	.55	.70	.75
4 "	.35	55	.70	.85	.90

Sink Strainers.

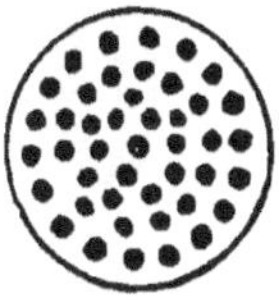

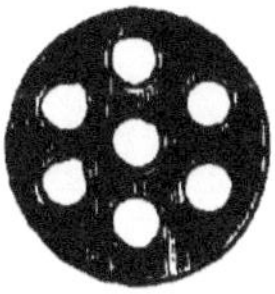

SIZE	¾	1	1¼	1½	2	2½	3	3½	4 in.
Small Holes, per doz. / Large " "	.60	.85	1.20	1.45	1.80	2.40	3 00	3.60	4.80
Fancy " " / Convex " "	.96	1.44	1.80	2.40	3.00	3.60	4.20	4 80	6.00

Bell Traps.

SIZE	2½	3	3½ in.
Lead, per dozen	9.00	12.00	14.00

Water Back Couplings.

Straight.

SIZE	½	¾	1 in.
Plain Face, per dozen	6.50	7.00	10.00
Ground " "	7.50	8.00	11.50

Water Back Couplings.

Bent.

SIZE	½	¾	1 in.
Plain Face, per dozen	7.50	8.00	11.00
Ground " "	8.50	9.00	12.50

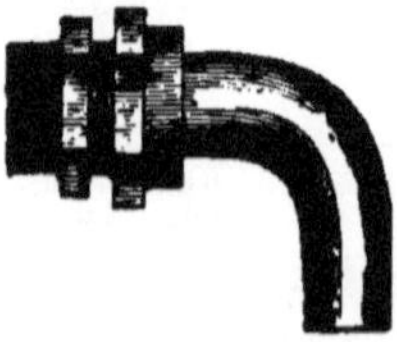

Boiler Coulpings.

Bent, for Iron Boilers.

SIZE	½	¾	1 in.
Plain Face, per dozen	8.50	9.00	12.00
Ground " "	9 50	10.00	13.50
" " per set, 3 Bent, 1 Straight..	......	2.75	....
Plain " " 3 " 1 "	......	2.50	....

Boiler Couplings.

Straight, for Iron Boilers.

SIZE	½	¾	1 in.
Plain Face, per dozen	7.50	8.00	11.00
Ground " "	8 50	9.00	12.50

Boiler Couplings.

Bent, for Iron Pipe.

SIZE	½	¾	1 in.
Ground Joint, per dozen	10.50	11.00	14.50
Plain " "	9.50	10.00	13 00

Boiler Couplings.

Straight, for Iron Pipe.

SIZE	½	¾	1 in.
Ground Joint, per dozen	9.50	10.00	13.50
Plain " "	8 50	9.00	12.00
Per set, 3 Bent, 1 Straight, Ground Joint	......	3.00	......
" " 3 " 1 " Plain "	......	2.75	......

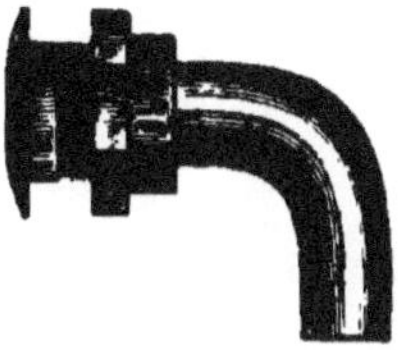

Boiler Couplings.

Bent, for Copper Boilers.

SIZE	½	¾	1 in.
Plain Face, per dozen	8.00	8 50	11.00
Ground " "	9.00	9.50	12.50

Boiler Couplings.

Straight, for Copper Boilers.

SIZE	½	¾	1 in.
Plain Face, per dozen	7.00	7.50	10.00
Ground " "	8.00	8.50	11.50
" " per set,	......	2.75	....
Plain " "	......	2.50	....

Coupling for Brown Bros. Boilers.

Per set of 4 .... $2.50

Plain Coupling,

for Lead Pipe.

SIZE	¼	⅜	½	⅝	¾	1	1¼	1½	1¾	2 in.
Plain Face, dozen.	3.00	3.50	4.00	5.00	6.50	10.00	15.00	2 .00	28.00	30.00
Ground Face, "	4.00	4.50	5.00	6.50	8.00	12.00	18.00	24.00	33.00	36.00

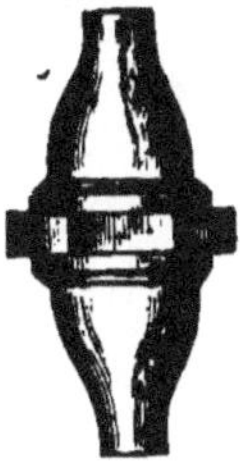

Valve Couplings.

Ground Joint.

SIZE	½	⅝	¾	1	1¼	1½	2 in.
For Lead Pipe, per dozen	10.00	12.00	15.00	20.00	30.00	40.00	60.00
" Iron " "	11.00	14.00	17.00	23.00	34.00	44.00	70.00

Soldering Nipples.

Male or Female.

Size...	1/4	3/8	1/2	3/4	1	1 1/4	1 1/2	2	2 1/2	3 in.
Per dozen...	1.75	2.25	2.50	3.00	5.00	7.50	10.00	14.00	20.00	28.00

Soldering Unions.

Size...	1/4	3/8	1/2	3/4	1	1 1/4	1 1/2	2	2 1/2	3 in.
Per dozen..	2.25	2.75	3.25	4.00	6.00	8.50	12.00	18.00	23.00	33.00

Cistern Valves.

Size.................................	1 1/4	1 1/2	2 in.
A, per dozen.................................	12.00	12.00	....
B, " Small............................	....	....	20 00
" " Large............................	....	....	25 00

Closet Cranks.

Horizontal or Upright, per dozen...............................$3.00

Ball Lever.

Per dozen $6.00

Street Washer Screws.

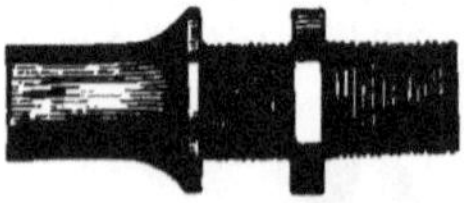

Size..	1/2	3/4	1 in.
Short, for Iron Pipe, per dozen..................................	5.00	6.00	7.50
Long, for Lead Pipe, "	5.50	6.00	8.50
Long, for Iron Pipe, "	6.00	7.00	10.00
With Flange and Cap, for Lead Pipe, per dozen..................	9.00	9.00	....
With Flange and Cap, for Iron Pipe, "	9.00	9.00	....

Street Washer Checks.

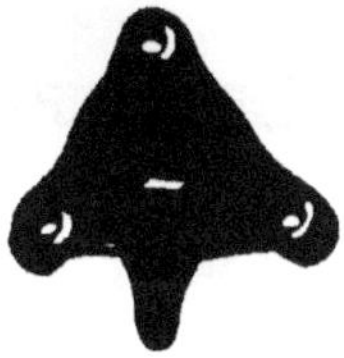

Per dozen $3.00

Hydrant Nozzles.

SIZE	½	¾	1 in.
For Lead Pipe, per dozen	6.00	7.50	9.00
" Iron " "	8.00	9.50	11.00

Hydrant Handles.

Plain, with Guide,	per dozen	$7.00
For Compression Cocks,	"	5.00
With Triangular Guide,	"	8.00

Hydrant Clamps.

Brass,	Small,	per dozen	$2.50
"	Large,	"	3.50
Iron,	Small,	"	1.50
"	Large,	"	2.25

Hydrant Sockets.

Brass, per dozen $2.00

Iron Street Washer Key.

Each .... $ .30

Iron Street Washer Rod.

Each .... $ .50 Hydrant Rod, each .... $ .75

Hose Nozzles.

To Tie On.

SIZE	½	¾	1 in.
Length	$4\frac{1}{2}$	$4\frac{1}{2}$	$4\frac{1}{2}$
Per dozen	3.00	3.50	4.00

Hose Pipes.

Without Tip.

Size	3/4	1	1¼	1½	2	2½ in.
Length	8	8	12	12	12	15 in.
Per dozen	7.00	9.00	18.00	22.00	34.00	65.00

Hose Pipes.

Plain, Cast, Screw Tip.

Size	3/4	3/4	3/4	1	1	1¼	1¼	1½	1½	2	2	2½ in
Length	6	8	12	8	12	12	15	12	15	12	20	15
Per dozen	......	8.00	10.00	10.00	12.00	20.00	24.00	25.00	30.00	38.00	50.00	75.00

Hose Pipes.

Short, Cock Large End, Cast.

Size	3/4	3/4	3/4	3/4	1	1	1¼	1¼	1½	2 in.
Length	6	8	9	12	8	12	12	15	15	12
Per dozen	11.00	13.00	18.00	18.00	15.00	20.00	40.00	45.00	60.00	80.00

Hose Pipes.

Compression, Short, Cock Large End, Cast.

Size	3/4	3/4	1	1 in.
Length	8	12	8	12
Per dozen	20.00	24.00	24.00	28.00

Hose Pipe Tips.

Size	3/4	1 in.
Price, per dozen	4.00	4.00

Combination Hose Pipe.

Throws a Solid Stream or Spray.

"The Daisy," Brass.		per dozen	$12.00
"	Nickel Plated,	"	14.00

Boss Hose Pipe.

Size	¾	1
Nickeled, per dozen	12.00	14.00

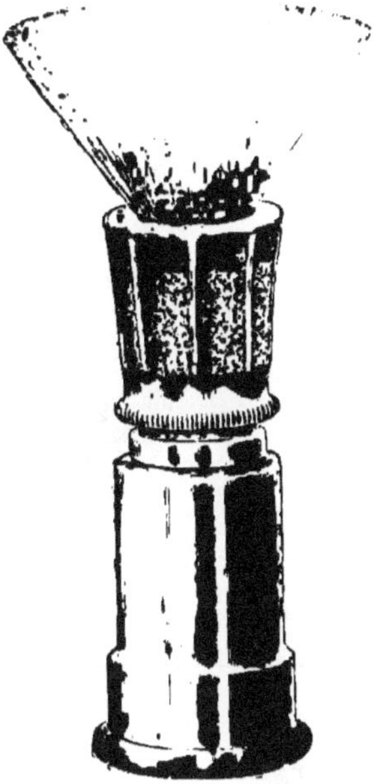

Crown Hose Pipe.

Size	¾
Nickeled, per dozen	10 00

Hose Sprinklers.

Size	1½	2	2½	3	3½	4 in.
Per dozen	3.50	4.50	6.00	9 00	12.00	18.00

Hose Couplings.

For Rubber Hose.

Size	½	¾	1	1¼	1½	2	2½	3 in
Per dozen	2.40	2.40	4.40	10.00	14.00	30.00	48.00	76.00
" For Iron Pipe	2.65	2.65	4.65	10.50	15.00	32.00	50.00	79.00

For either part of Coupling two-thirds List Price.

Hose Splices.

For Mending Hose.

SIZE	½	¾	1 in.
Brass, per dozen	1.20	1.20	2.00
Iron, Coppered, "	.40	.50	1.00
All Iron	.40	.50	.85

Hose and Iron Pipe Nipples.

SIZE	¾	1	1¼	1½	2	2½	3 in.
Per dozen	3.50	5.00	9.00	10.00	14.00	28.00	40.00

Hose Reducers.

SIZE	1 x ¾	1¼ x 1	1½ x 1¼	2 x 2½
Price, per dozen	6.50	10.00	12.00	18.00

Hose Bushing.

1 x ¾, per dozen $6.50

Hose Bibb Ends.

SIZE	½	⅝	¾	1	1¼	1½	2 in.
Per dozen	2.50	2.50	2.50	3.50	6.00	8.00	15.00

Hose Caps.

Size	3/4	1	1 1/4	1 1/2	2	2 1/2 in.
Per dozen	4.00	6.00	8.00	10.00	15.00	24.00

The Caldwell Patent Hose Strap.

	Per dozen.		Per dozen.
No. 2, 1/2 inch, 3 3/8 inch long, " 4, 1/2 " 3 3/4 "	$ .40	No. 18, 1 1/2 inch, 6 3/4 inch long, " 20, 1 1/2 " 7 1/8 "	$1.20
No. 6, 3/4 inch, 4 1/8 inch long, " 8, 3/4 " 4 3/4 "	.60	No. 22, 1 3/4 inch, 7 1/2 inch long, " 24, 1 3/4 " 8 "	1.40
No. 10, 1 inch, 5 inch long, " 12, 1 " 5 3/8 "	.80	No. 26, 2 inch, 8 1/2 inch long, " 28, 2 " 9 "	1.60
No. 14, 1 1/4 inch, 6 inch long, " 16, 1 1/4 " 6 3/8 "	1.00	No. 30, 2 1/4 inch, 9 1/2 inch long, " 32, 2 1/4 " 10 "	1.80
		No. 34, 2 1/2 inch, 10 1/2 inch long, " 36, 2 1/2 " 11 "	2.00

Hose Strap Fasteners or Plyers.

1/2 to 1 inch, inclusive	$ .50
1 1/4 " 2 1/2 " "	.75

Improved Double Hose Band.

Hydrant Hose.

Size of Hose in In.—3 ply	1/2	3/4	1	1 1/4	1 1/2	1 3/4	2	2 1/4	2 1/2	2 3/4	3 in.
Price of Bands, per doz.	1.50	1.50	2.00	2.50	3.00	3.50	4 00	5.50	7.00	8.50	10.00

Steam and Brewers' Hose.

Price of Bands, per dozen	1.50	2.00	2.50	3.00	3.50	4.00	5.50	7.00	8.50	10.00	13.00

Lawn Sprinklers.

Complete, with Stand.

Four Arm each	$6.50
Eight Arm, "	8.00

We also furnish various other makes of Lawn Sprinklers. Circulars furnished on application.

Chain.

Safety.

NUMBERS	20	19	18	17	16	15	14	13
Single, Brass, per doz. yds.	.70	.85	1.00	1.10	1.25	1.40	1.70	2.05
" Plated, " "	1.60	2.00	2.25	2.50	2.75	3.00	3 25	4.00
Double, Brass, " "	1.40	1.60	1.90	1.95	2.25	2.55	3.00	4.05
" Plated, " "	2.40	2.70	3.10	3.25	3.75	4.25	5.00	6.50

Plumbers' Safety Chains.

Put Up 12 Yards in a Box.

No. 0,	Brass,	Dipped and Lacquered,	per box	$2.25
" 0,	"	Silvered,	"	2.50
" 0,	"	Nickel Plated,	"	2.75
" 1,	"	Dipped and Lacquered,	"	3.00
" 1,	"	Silvered,	"	3.25
" 1,	"	Nickel Plated,	"	3.50
" 2,	"	Dipped and Lacquered,	"	3.75
" 2,	"	Silvered,	"	4.00
" 2,	"	Nickel Plated,	"	4.25

Plumbers' Safety Chains on Reels.

500 Feet on a Reel.

No. 0, Brass	Per 100 feet,	$6.25
" 1, "	" " "	8.30
" 2, "	" " "	10.40

☞Nickel Plating, 25 Cents per 100 feet extra, *Net.*
Silvering, 15 cents per 100 feet extra, *Net.*

PLUMBERS' SAFETY CHAINS.

For Basins, Wash Trays and Bath Tubs.

WITH SPLIT CONNECTING LINKS ATTACHED TO EACH END.

EACH STYLE PUT UP 2 DOZEN IN A BOX.

Basin Chains.

16 Inches Long.

No. 0,	Brass,	Dipped and Lacquered,	per dozen	$1.00
" 0,	"	Silvered,	"	1.35
" 0,	"	Nickel Plated,	"	1.70
" 1,	"	Dipped and Lacquered,	"	1.10
" 1,	"	Silvered,	"	1.45
" 1,	"	Nickel Plated,	"	1.80

Wash Tray Chains.

20 Inches Long.

No. 0,	Brass,	Dipped and Lacquered,	per dozen	$1.15
" 0,	"	Silvered,	"	1.60
" 0,	"	Nickel plated,	"	2.00
" 1,	"	Dipped and Lacquered,	"	1.25
" 1,	"	Silvered,	"	1.70
" 1,	"	Nickel Plated,	"	2.10

Bath Tub Chain.

26 Inches Long.

No. 0,	Brass,	Dipped and Lacquered,	per dozen	$1.25
" 0,	"	Silvered,	"	1.75
" 0,	"	Nickel Plated	"	2.20
" 1,	"	Dipped and Lacquered,	"	1.40
" 1,	"	Silvered,	"	1.90
" 1,	"	Nickel Plated		2.35

Split Links For Safety Chain.

No. 2,	Brass,	Ordinary Size,	per gross	$2.00
" 2,	Nickeled,	" "	"	2.10
" 1,	Brass,	Large "	"	2.25
" 1,	Nickeled,	" "	"	2.35

Round Head Plated Screws.

1 inch, per doz....$ .14	1½ inch, per doz....$ .20	1¾ inch, per doz....$ .24

Closet Screws and Washers.

Brass, Nickel Plated, Solid Hexagon Head,	per 100	$4.00
" " " Loose " "	"	7.50
" " " Washers, Round,	"	1.00
" " " " Eliptical,	"	1.50

Patent Cushion Pipe Band.

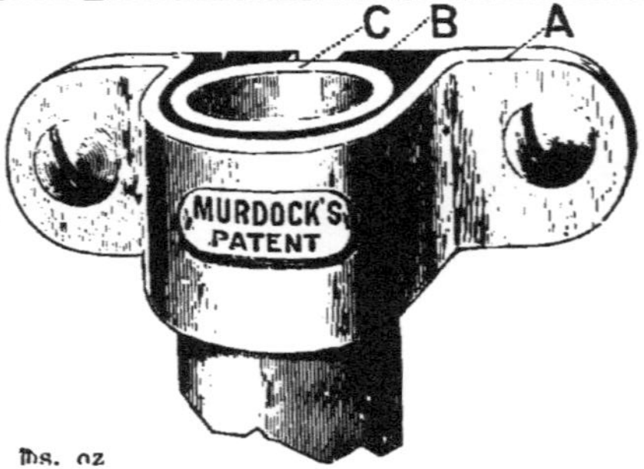

PRICE LIST.

				lbs. oz	Per Dozen. Brass.	Per Dozen. Nickel Plated.
No. 1A.	Band marked ⅜ X, with Cushion marked ½ L,	Fi s These Sizes Pipes:	⅜ Strong Lead Pipe, ½ Extra Light " ½ Light " ½ Medium "	(A 1 : 2 per foot) (D 0 : 12 " (C 1 : 0 " (B 1 : 3 "	$1.50	$1.70
No. 1B.	Band marked ⅜ X, with Cushion marked ⅜ X,	Fits These Sizes Pipes:	⅜ Extra Strong Lead Pipe, ½ Strong "	(AA 1 : 5 per foot.) (A 1 : 10 "	$1.50	$1.70
No. 2.	Band marked ½ X, with Cushion marked ½ X,	Fits These Sizes Pipes:	⅜ XX Strong Lead Pipe, ½ Extra Strong "	(AAA 1 : 12 per foot,) (AA 2 : 0 "	$2.25	$2.50
No. 3A.	Band marked ⅝ X, with Cushion marked ½ XX,	Fits These Sizes Pipes:	½ XX Strong Lead Pipe, ⅝ Strong " ⅝ Medium " ⅝ Light " ⅝ Extra Light "	(AAA 3 : 0 per foot,) (A 2 : 8 " (B 2 : 0 " (C 1 : 7 " (D 1 : 4 "	$2.25	$2.50
No. 3B.	Band marked ⅝ X, with Cushion marked ⅝ X,	Fits These Sizes Pipes:	⅝ Extra Strong Lead Pipe, ¾ Extra Light " ¾ Light "	(AA 2 : 12 per foot,) (D 1 : 3 " (C 1 : 12 "	$2.25	$2.50
No. 3C.	Band marked ⅝ X, with Cushion marked ⅝ XX,	Fits These Sizes Pipes:	⅝ XX Strong Lead Pipe, ¾ Strong " ¾ Medium "	(AAA 3 : 8 per foot.) (A 3 : 0 " (B 2 : 3 "	$2.25	$2.50
No. 4A.	Band marked ¾ X, with Cushion marked ¾ XX,	Fits These Sizes Pipes:	¾ XX Strong Lead Pipe, 1 in. Light " 1 in. Extra Light "	(AAA 4 : 0 per foot.) (C 2 : 8 " (D 2 : 0 "	$2.50	$2.75
No. 4B.	Same Band, with Cushion, ¾ X,	Fits This Size Pipe:	¾ X Strong	(AA 3 : 8 per foot,)	$2.50	$2.75
No. 5.	Band marked 1 X Strong with Cushion marked 1 in. X Strong.	Fits These Sizes Pipes:	1 in. Medium Lead Pipe, 1 in. Strong " 1 in. Extra Strong " 1 in. XX Strong " 1¼ in. Extra Light " 1¼ in Light "	(B 3 : 4 per foot,) (A 4 : 0 " (AA 4 : 8 " (AAA 6 : 8 " (D 2 : 8 " (C 3 : 0 "	$3.00	$3.40

Above prices include screws.

Repairs for Compression Bibbs.

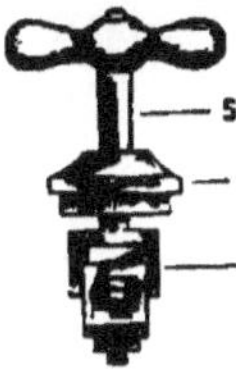

SIZE	3/8	1/2	5/8	3/4	1 in.	
Handles, per doz.	1.25	1.50	1.75	2.50	4.50	....
Spindles, "	1.25	1.50	1.75	2.50	4.50	....
Caps, "	1.25	1.50	1.75	2.50	4.50	....
Valves, "	1.25	1.50	1.75	2.50	4.50	....

Packings and Washers for Compression Bibbs.

SIZE	3/8	1/2	5/8	3/4	1 in.
Common Rubber for Cup, per 100	.75	.75	.75	1.00	1.50
" " " Valve, " "	.40	.40	40	.50	.65
Black " " " " "	1 25	1.25	1.38	1.50	2.00
Vulcanized Fibre, " " " "	.50	.50	.50	.60	.75
Jenkins' " " " "	.75	.75	.75	.75	.75
Boss Washers, " " " "	1.00	1.00	1.00	1.00	1.50

Washers.

3/4 inch Leather Hose Coupling Washers, per 100	$ .50
1 " " " " " " "	.60
Pasteboard Boiler Washers, " "	.30

Pump Leathers.

NUMBER	1	2	3	4
Valves, per dozen	.50	.50	.60	.70
Plungers, "	.50	.50	.60	.70

Pump Leather for valves in the side, uncut, per lb., 45 cents.

Plumbers' Prepared Soil.

Quart Cans, each, net,	$ .35
Pint " " "	.20

Gas Fitters' Cement.

Per pound, net	$ .15

Steam Joint Cement.

Steam Joint Cement, per pound, net	$ .20
Rustless Steam Joint Cement, per pound, net	.15

Copper Bath Tubs.

Regular.

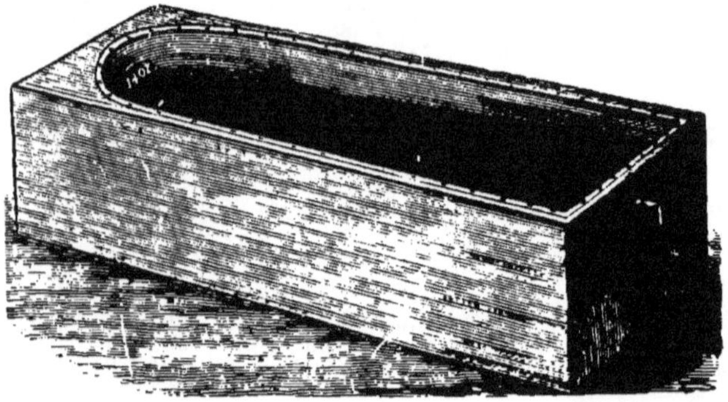

SIZES: 5, 5½ or 6 feet long; 24 in. wide, 19½ in. deep, outside measure.

Weight of Copper to sq. ft.	8	10	12	14	16	18	20 oz.
Each	12.75	15.00	16.00	18.00	20.00	22.00	24.00

Copper Bath Tubs.

French Pattern.

26 in. wide at top, 23 in. wide at bottom, 22 in. deep, outside measure.

Weight of Copper to square foot.	8	10	12	14	16	18	20 oz.
4½ feet long	14.00	16.00	17.00	19.00	21.00	23.00	25.00
5 " "	......	18.00	19.00	21.00	23.00	25.00	27.00
5½ " "	......	20.00	21.00	23.00	25.00	27.00	29.00
6 " "	......	22.00	23.00	25.00	27.00	29.00	31.00
Copper Hip or Seat Baths	......	10.00	11.00	12.00	13 00	14.00	15.00
" Foot "	......	7.50	8.50	9.50	10.50	11.50	12.50

Zinc Bath Tubs.

Any Size up to 6 feet long.

Each, Dovetailed case, with overflow	$8.00
" Nailed " open head, short bottom, no overflow	7.00

Steel Clad Bath Tubs.

French Pattern.

DIMENSIONS: 5 feet 3 inches, 5 feet 9 inches, and 6 feet by 26 inches wide outside of rim; 20 inches deep. Height from floor to top of rim, 26 inches.

Weight of Copper	12	14	16 oz.
Number	20	22	24
Size: 5 ft. 3 in., Price	31.00	33.00	35.00
Number	26	28	30
Size: 5 ft. 9 in., Price	31.50	33.50	35.50
Number	32	34	36
Size: 6 ft., Price	33.00	35 00	37.00

Roman Pattern.

DIMENSIONS: 4 feet 6 inches, and 5 feet by 26 inches outside of rim; 20 inches deep. Height from floor to top of rim, 26 inches.

Weight of Copper	12	14	16 oz.
Number	21	23	25
Size: 4 ft. 6 in., Price	31.50	33.50	35.50
Number	27	29	31
Size: 5 ft., Price	33.50	35.50	37.50

We also furnish to order, **Porcelain Lined and Enameled Iron Bath Tubs.**

Copper Sinks.

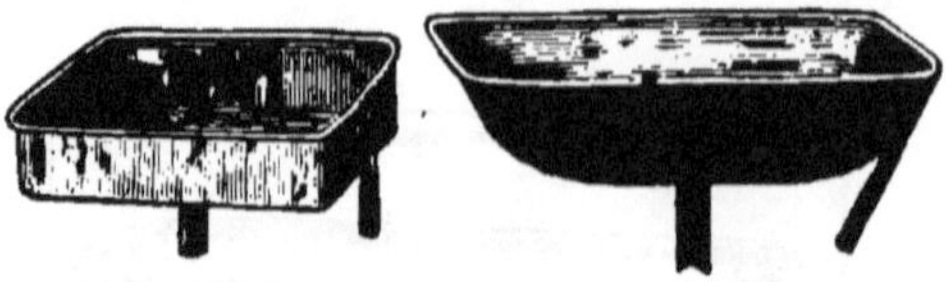

SIZE	12x18	12x20	14x16	14x20	14x24	16x24	16x30	18x30 in.
Square Bottoms	4.50	5.00	4.50	6.00	7.00	8.00	10.00	11.00
Oval "	6.00	6.50	6.00	7.50	9.00	10.00	12.00	13.00

Copper Wash Basins.

SIZE, inside measure	12	14 in.
Price, per dozen	4.00	5.00

Copper Closet Pan.

Light,	per dozen	$7.00
Extra Heavy,	"	8.00
Double Weight,	"	12.00

Copper Balls.

SIZE	4	5	6	7	8	10	12 in.
Price, per dozen	4.50	6.00	7.00	10.50	80c. per pound.		

Copper Showers.

No. 1, Plain Tinned,	per dozen	$12.00
" 2, Fancy,	"	15.00
" 3, " Flange and Thimble,	"	17.50

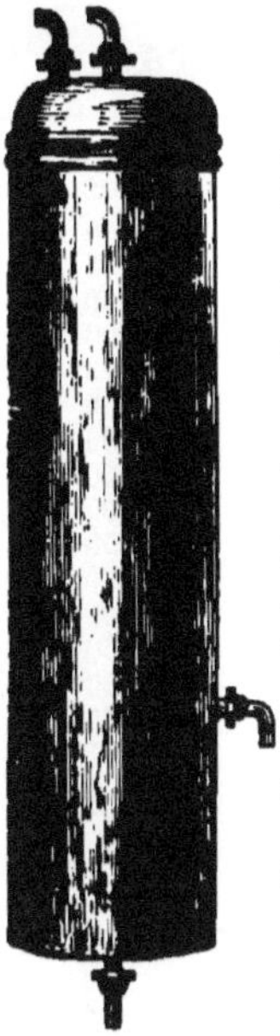

Copper Boilers.

No. of Gallons.....	30	35	40	45	50	60	100
Light Pressure, Flat Head,	22.00	24.00	28.00	35.00	38.00	48.00	84.00
Heavy Pressure, Flat Head,	24.00	27.00	32.00	37.00	41.00	52.00	88.00
Light Pressure, Round Head,	24.00	27.00	32.00	37.00	41.00	52.00	88.00
Heavy Pressure, Round Head,	26.00	30.00	34.00	39.00	43.00	55.00	92.00
Double Pressure, RoundHead.	....	....	....	....	......	85.00	112.00
Boxing...............	1.25	1.25	1.50	1.50	1.75	2.00	....

Above Prices Include Couplings and Tube.

The Light Pressure Boilers are intended for Tank Pressure only.

The Heavy Pressure Boilers are intended for the heavier pressure of the Water Works System.

Copper Boilers should be fitted with Vacuum Valves.

Brown Bros. Seamless Copper Boiler.

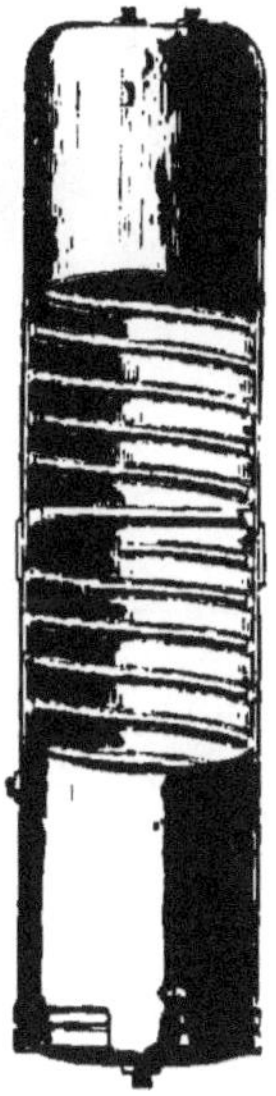

Capacity, Gallons....	30	35	40	45	50	60
Regular Pressure.....	30.00	35.00	40.00	45.00	50.00	60.00
Ext. Heavy Pressure,	40.00	45.00	50.00	55.00	65.00	80.00
Boxing net...........	1.00	1.00	1.25	1.25	1.50	1.50

These boilers are guaranteed to stand a Vacuum and are all tested. Regular pressure 200 lbs. per square inch. Extra Heavy Pressure 300 lbs. per square inch. They are made of the best quality of copper, and are thoroughly tinned on the inside.

Galvanized Iron Boilers

SIZE.	CAPACITY.	BLACK OR GALVANIZED.
3 feetx12 inch.	18 Gallons	$ 14.50
4 " x12 "	24 "	15.75
4½ " x12 "	27 "	18.50
5 " x12 "	30 "	19.00
4 " x14 "	32 "	21.00
5 " x13 "	35 "	21.00
5 " x14 "	40 "	24.00
4 " x16 "	42 "	26 00
5 " x16 "	52 "	31.00
4 " x18 "	53 "	31.50
6 " x16 "	63 "	38.00
5 " x18 "	66 "	38.00
5 " x20 "	82 "	45.50
5 " x22 "	100 "	63.50
5 " x24 "	120 "	72.50
6 " x24 "	144 "	103.00
8 " x24 "	192 "	132.00

Coupling and Tube extra.

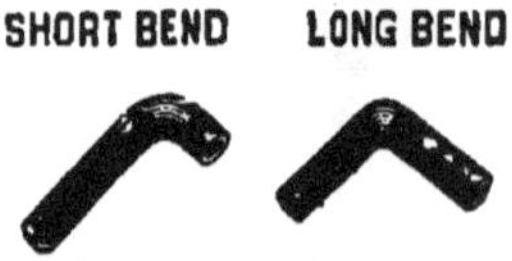

Lead Traps.

"Du Bois."

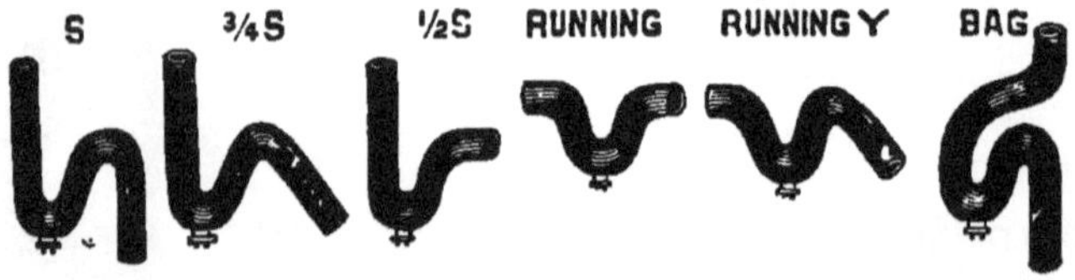

Shapes	Extra Heavy					Special	Standard Weight					
	8 lb. Lead				6 lb.	5½ lb.	6 lb. Lead				5 lb.	4½ lb
	4½	4	3	2	1½	4	4½	4	3	2	1½	1¼ in.
Full S	3.60	2.85	2.15	1.30	.70	1.90	2.80	2.20	1.70	.95	.60	.50
¾ S	3.60	2.85	2.15	1.30	.70	1.90	2.80	2.20	1.70	.95	.60	.50
Half S	2.75	2.15	1.75	1.00	.55	1.45	2.10	1.65	1.30	.80	.50	.40
Running	2.75	2.15	1.75	1.00	.55	1.45	2.10	1.65	1.30	.80	.50	.40
Running Y	3.60	2.85	2.15	1.30	.70	1.90	2.80	2.20	1.70	.95	.60	.50
Bag	5.15	4.25	2.80	1.65	.90	2.75	3.90	3.20	2.15	1.25	.80	.65
Short Bend	1.65	1.35	.85	.60	.32	.88	1.25	.97	.65	.41	.30	.20
Long Bend	2.10	1.65	1.20	.80	.41	1.10	1.60	1.25	.90	.55	.36	.26

The above prices include Brass Drain Screws for all Traps, excepting the 4½-inch and 4 inch sizes.

N. B. Standard Weight only kept in Stock, Extra Heavy and Special furnished to order.

Vented Traps.

WITH BRASS VENT CONNECTIONS.

All have Drain Screws at Bottom.

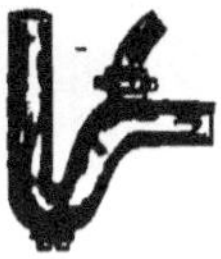

PRICE LIST.

	1¼	1½	2	
			1½ Vent	2 Vent
Full S	.90	1.10	1.45	1.65
¾ S	.90	1.10	1.45	1.65
P	.80	1.00	1.30	1.50
Running	.80	1.00	1.30	1.50
Extra Vent Connections	.30	.40	.40	.60
Extra Tail Pieces only	.15	.20	.20	.30

Extra Long Traps.

PRICE LIST.

	1¼	1½	4 in.
Full S	.80	1.00	......
" Vented	1.20	1.50	......
P	......	......	1.80

Compressed Lead Traps.

With Plain Screw Cap.

Screw Cap with Straight Vent.

Screw Cap with Elbow Vent.

Outside Dimensions of Traps.

3 inch Diameter, with Plain Screw Cap, each	$1.60
4 " " " " " " "	2.00
5 " " " " " " "	2.75
6 " " " " " " "	3.75
3½ inch Diameter, with Straight or Elbow Vented Cap, each	2.35
4 " " " " " " " " "	2.75
5 " " " " " " " " "	3.60
6 " " " " " " " " "	4.75

PRICES OF VENTED TRAP SCREW TOPS.

3½ inch Vented Trap Screw, either Straight or with Elbow	$ .65
4 " " " " " " " " "	.75

Star Traps.

1¼ Inch Water Seal.	1¼	1½	2 in.	3 Inch Water Seal.	1¼	1½	2 in.
Full S	$0.45	$0.50	$0.60	Full S	$0.55	$0.55	$0.95
¾ S	.45	.50	.....	¾ S	.55	.55	.95
Half S	.40	.45	.50	Half S	.50	.50	.85
Running........	.....	.65	.....	Running.......	.....	.65	.....

Improved Star Traps.

	1½ in.
Full S..	$0 65
¾ S..	.65
Half S	.60
Running..	.75

Running Star Trap.

No. 6, with Connection for Bath...............$1.00

Cudell Traps.

One price for Full S, ½ S, ¾ S, Running and Bath Traps. Prices of Trap with full or open cover and Vent dome correspond.

SIZE { Inlet.................................. / Outlet..................................	1 / 1¼	1¼ / 1¼	1¼ / 1½	1½ / 1½	1½ in. / 2 in.
Lead Body, Hard Metal Dome and Cover.......	1.00	1.00	1.25	1.25	1.35
" " " " " " Vent Top...	1.25	1.25	1.50	1.50	1.60
Lead Body, Hard Metal Dome, Vent Top and Union....................	1.50	1.50	1.75	1.75	1.85
Lead Body, Brass Dome and Cover....	1.50	1.50	1.75	1.75	1.85
" " " " " Vent Top..........	1.75	1.75	2.15	2.15	2.25
" " " " Vent Top and Union..	2.25	2.25	2.65	2.65	2.75

PRICE LIST.

Extras for Traps 1 x 1¼ in. to 1½ x 2 in. inclusive.

Vent Top Hard Metal Coupling,	per dozen	$4.80
" " Brass Coupling,	"	7.20
" " Union Hard Metal Coupling,	"	3.00
" " " Brass "	"	6.00
Full Cover, Hard Metal,	"	1.80
" " Brass,	"	4.20
Open " Hard Metal,	"	1.80
" " Brass,	"	4.20
Trap Balls,	"	1.80
Rubber Washers,	"	.40

Bower Traps.

Regular Styles.

Form.		Full S.				
TRAP NUMBER		1	5	11	15	21
SIZE { Inlet		1	1¼	1¼	1½	1½ in.
SIZE { Outlet		1¼	1¼	1½	1½	2 in.
	Style.	Price.	Price.	Price.	Price.	Price.
Lead and Glass	A	1.00	1.05	1.10	1.38	1.50
All Lead	B	1.20	1.25	1.30	1.63	1.75
Lead and Brass	C	1.40	1.45	1.50	1.88	2.00
Lead and Brass, N. P.	D	1.50	1.55	1.60	2.00	2.12
Brass Body Screw and Glass	E	1.45	1.50	1.55	1.88	2.00
" " " " Lead	F	1.65	1.70	1.75	2.13	2.25
" " " " Brass	G	1.85	1.90	1 95	2.38	2.50
" " "n.p." Glass	H	1.50	1.55	1.60	1.94	2.06
" " " " Brass n. p	I	2.00	2.05	2.10	2.58	2.70

Form.		Half S.				
TRAP NUMBER		3	7	13	17	23
SIZE { Inlet		1	1¼	1¼	1½	1½ in.
SIZE { Outlet		1¼	1¼	1½	1½	2 in.
	Style.	Price.	Price.	Price.	Price.	Price.
Lead and Glass	A	.95	1 00	1.05	1.30	1.40
All Lead	B	1.15	1.20	1.25	1.55	1.65
Lead and Brass	C	1.35	1.40	1.45	1.80	1.90
Lead and Brass, N. P.	D	1.45	1.50	1.55	1.92	2.02
Brass Body, Screw and Glass	E	1.40	1.45	1.50	1.80	1.90
" " " " Lead	F	1.60	1.65	1.70	2.05	2.15
" " " " Brass	G	1.80	1.85	1.90	2.30	2.40
" " "n.p." Glass	H	1.45	1.50	1.55	1.86	1.96
" " " " Brass, n.p.	I	1.95	2.00	2 05	2.50	2.60

Form.		Running.	
TRAP NUMBER		9	10
SIZE { Inlet		1¼	1½ in.
SIZE { Outlet		1¼	1½ in.
	Style.	Price.	Price.
Lead and Glass	A	1.10	1.40
All Lead	B	1.30	1.65
Lead and Brass	C	1.50	1.90
Lead and Brass, N. P.	D	1.60	2.02
Brass Body, Screw and Glass	E	1.55	1.90
" " " " Lead	F	1.75	2.15
" " " " Brass	G	1.95	2.40
" " "n.p." Glass	H	1.60	1.96
" " " " Brass, n. p.	I	2.10	2.60

N. B.—The Trap Number indicates the size and form. The Letter indicates the style. N. P. signifies that the brass portion of trap is nickel-plated.

In ordering be sure and give number and letter or size, form and style.

Where size and form are given without style, we send the Lead and Glass.

In the above list of Traps with Brass Cups prices are given for Sheet Brass Cups. If Cast Brass Cups are wanted, add to the price the difference between the Cast and Sheet Brass Cups, as given in the list of extras.

PRICE OF EXTRAS.

Small Glass Cups,		(for Traps 1 x1¼ to 1¼x1½,	inclusive) each	$ .10
" Lead "		" 1 x1¼ " 1¼x1½,	" "	.30
" Sheet Brass "		" 1 x1¼ " 1¼x1½,	" "	.50
" " " "	N. P	" 1 x1¼ " 1¼x1½,	" "	.60
" Cast " "	——	" 1 x1¼ " 1¼x1½,	" "	.70
" " " "	N. P.	" 1 x1¼ " 1¼x1½,	" "	.80
Large Glass "		" 1½x1½ and 1½x2,	" "	.15
" Lead " "		" 1½x1½ " 1½x2,	" "	.40
" Sheet Brass "		" 1½x1½ " 1½x2,	" "	.65
" " " "	N. P.	" 1½x1½ " 1½x2,	" "	.75
" Cast " "	——	" 1½x1½ " 1½x2,	" "	.90
" " " "	N. P.	" 1½x1½ " 1½x2,	" "	1.00
Extra Large Sheet Brass Cups,		for 2 inch Universal	"	.85
" " " " "	N. P.	" 2 " "	"	.95
" " Cast " "	——	" 2 " "	"	1.10
" " " " "	N. P.	" 2 " "	"	1.20

Different Styles of Cups are Interchangeable—corresponding sizes fitting any of our Traps, Combination as well as Regular Styles.

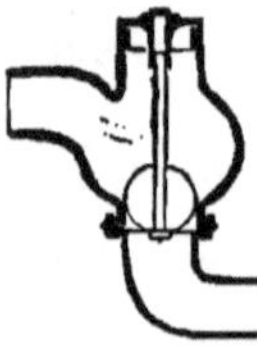

The Delehanty Rough Brass Bath Trap

Without Back Air Vent.

Price each, 1½ in.	$ 3.00
" 2 " with connections for Iron Bath Tubs	10.00

" For above Bath Trap with back air connections, add to List, 1½ in., $1.50 ; 2 in., $2.00.

McClelland's Anti-Syphon Vent Trap.

Price each, 1¼ in	$2.25

This Cut Represents our Nickel Plated Trap

Connected to the ordinary patent overflow bowl in full S style with Back Air Connections.

PRICE LIST FOR VENTED, FULL OR ¾ S.

1¼ inch, Nickel Plated	$ 9.00
1½ " "	11.00
1¼ " Lacquered Brass	8.00
1½ " "	10.00

PRICES WITHOUT VENT, FOR FULL OR ¾ S.

1¼ inch, Nickel Plated	$ 8.00
1½ " "	10.00
1¼ " Lacquered Brass	8.00
1½ " "	10.00

We also furnish the above Traps fitted for Valve Waste attachments and common overflow Basins. Prices furnished on application.

N. B.—The foregoing styles of Lead Traps represent the kinds now most generally in use, but we can also furnish any other make of trap now on the market, and keeping pace with the times, will be prepared to supply the trade with any new device in this line that may be introduced.

PLUMBERS' EARTHEN-WARE.

Wash Basins.

OUTSIDE MEASURE	10	12	13	14	15	16 in.
Common Overflow, each	....	1.05	1.25	1.40	2.00	2.75
Without " "	....	.80	1.00	1.15	1.80	2.25

Patent Overflow Basins.

For Metal Plugs.

OUTSIDE MEASURE.	12	13	14	15	16 in.
Price, each	1.15	1.30	1.50	2.25	3.00

Patent Overflow Basins.

For Rubber Plugs.

OUTSIDE MEASURE	12	13	14	15	16 in.
Price, each	1.40	1.55	1.70	2.50	3.50

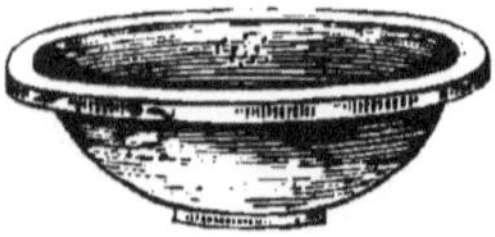

Oval Wash Basins.

SIZE	14x17	15x19
Without Overflow, each	3.50	5.00
Common " "	3.50	5.50
Patent " "	3.80	5.50
" " for Rubber Plug, "	4.25	5.00

Recessed Basins.

SIZE	14x17	15x19	SIZE	14	15	16 in.
Oval	7.00	8.50	Round	5.00	6.00	7.00

These prices do not include Standing Waste. Prices on above with Standing Waste on application.

French Round Closet Basins.

Price, each...$1.45

Oval Closet Bowls.

Price, each...$1.65

Closet Basins.

With Ventilator.

Round Pipe Wash and Ventilator, each...........$3.25
Oval " " " " " 3.70

Decorated Basins of all kinds and Designs,

Furnished to order.

Flat Bedfordshire Urinals.

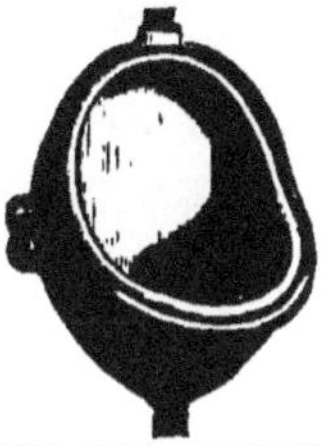

Number.....	3	2	1
Size........	Small. 11½x14	Medium. 12x15	Large. 15x18 in.
Each.............	6.70	7.10	9.30

Bedfordshire Corner Urinals

Number..	3	2	1
Size.....	Small. 10¼x10¼	Medium. 11x11	Large. 12x12 in.
Each......	6.70	7.10	9.30

Lip Urinals.

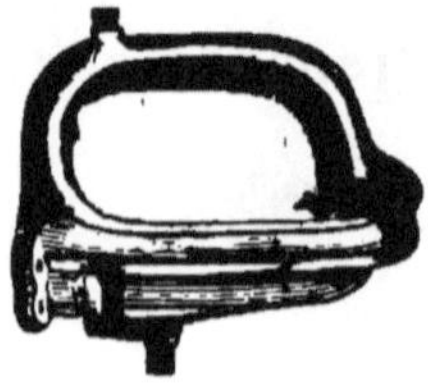

Number	3	2	1
SIZE, flat	Small. 12¼x14¼	Medium. 12x15	Large. 15x18 in.
Each	10.70	11 25	13.50
SIZE	10¼x10¼	11x11	12x12
Corner, each	10.70	11.75	13 50

Ventilated Urinals.

Flat Back or Corner with Hood and Lip.

Number	2	1
SIZE	12x15	15x18 in.
Each	13.90	17.45

Ohliger's Rubber Connection.

For connecting Waste Pipe to Rubber Plug Basins, Urinals, the overflow of common Overflow Basins, &c.

Per dozen $1.25

Coler's Patent Vulcanized Rubber Coupling.

SECTIONAL VIEW.

For Connecting Lead Pipe with Earthen Closet Basins.

Per dozen $1.50

Rubber Gaskets.

For Wash Basins.

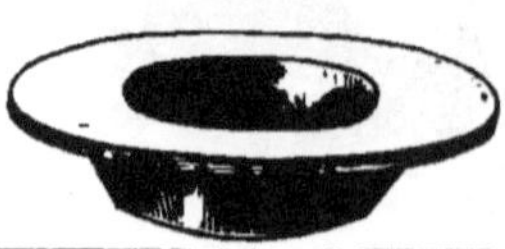

Price, per dozen $1.50

MARBLE SLABS.

All of these Slabs are 1¼ inch, countersunk, with ⅞ backs cut for 14 inch bowl, drilled for three Clamps, two Cock Holes and a raised place for Chain Stay Hole, unless specially ordered.

Customers ordering Slabs cut for one cock only, will be sent a chain stay suitable for other cock hole which is cut in slab, unless otherwise specified in ordering.

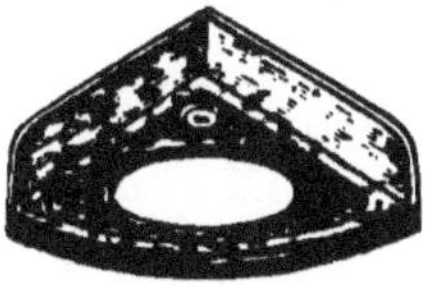

Corner Slabs.

Size of Marble	20x20	20x20	22x22	22x22	24x24 in.
Height of Back	8	10	8	10	10 in.
Size of Basin	14	14	14	14	14 in.
Contents of one	5 ft. 9 in.	6 ft. 4 in.	6 ft. 7 in.	7 ft. 3 in.	8 ft. 2 in.
Price, Italian, each	6.00	7.70	8.00	8.60	9.80

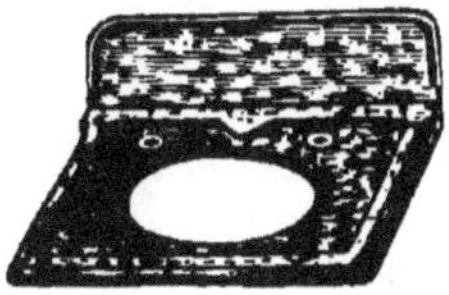

Single Back.

Size of Marble	20x24	20x27	20x27	20x30	20x30	20x33	20x36 in.
Height of Back	8	8	10	8	10	10	10 in.
Size of Basin	14	14	14	14	14	14	14 in.
Contents of one	5 ft. 5 in.	6 ft. 1 in.	6 ft. 6 in.	6 ft. 8 in.	7 ft. 2 in.	7 ft. 10 in.	8 ft. 6 in.
Price, Italian	6 60	7.30	7.80	8.00	8.60	9.40	10.20

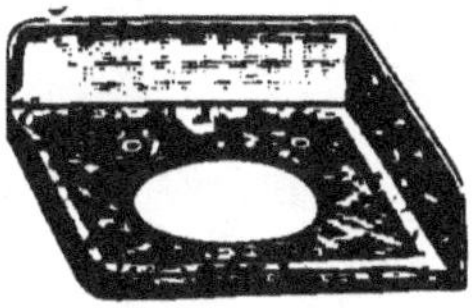

Back and Right Hand End.

Size of Marble	20x24	20x27	20x30	20x36 in.
Height of Back	8	10	10	10 in.
Size of Basin	14	14	14	14 in.
Contents of one	6 ft. 7 in.	7 ft. 10 in.	8 ft. 6 in.	9 ft. 11 in.
Price, Italian, each	7.90	9.40	10.20	11x90

Back and Left-hand End.

Size of Marble..................	20x24	20x27	20x30	20x36 in.
Height of Back..................	8	10	10	10 in.
Size of Basin..................	14	14	14	14 in.
Contents of one..................	6 ft. 7 in.	7 ft. 10 in.	8 ft. 6 in.	9 ft. 11 in.
Price, Italian, each..........	7.90	9.40	10.20	11.90

Slabs having more than one basin hole, will be furnished at the same price per foot, but 75 cents extra will be charged for each additional basin hole after the first.

Any size Marble Slabs made to order, but customers will oblige us by using above regular sizes as far as practicable.

Slabs 6½ ft. long and over will be charged 20 cts. per superficial ft. extra. We also furnish to order Recessed Slabs. Slabs cut for 13, 15 and 16 inches, Oval or Recessed Basins. Colored slabs of all sizes. Floor slabs for closets. Barber slabs and Drinking slabs of all designs. Urinal Stalls, (Marble and Slate.) Prices on all special slabs and Marble or Slate Urinal stalls, furnished on application.

All orders for Marble Slabs and Earthenware are accepted and filled by us at *buyers' risk only*, and under *no circumstances* can we assume any responsibility for safe delivery of such goods at destination.

Patent Imperishable Porcelain Wash-Tubs.

Clean! Impervious! Durable as Iron!

PRICE LIST.

Set of two Tubs, with Galvanished Iron Legs, and Ash Top Frame............	$36.00
Set of three Tubs, with Galvanised Iron Legs, and Ash Top Frame..........	52.00
Set of four Tubs, with Galvanized Iron Legs and Ash Top Frame..........	70.00

Vitrified Brown Glazed Earthenware Wash Tub.

PRICE LIST.

Set of two Tubs, with Galvanised Iron Legs, and Ash Top Frame............	$25.00
Set of three Tubs, with Galvanised Irot Legs, and Ash Top Frame............	37.50
Set of four Tubs with Galvanised Iron Legs, and Ash Top Frame............	50.00

Warranted capable of resisting the action of Acids and Alkalies, and to be absolutely Non-Absorbent.

Size of Each: Length, 27 inches; Width, 23 inches; Depth, 16 inches.

Granite Laundry Tubs.

With Patent Metallic Casing.

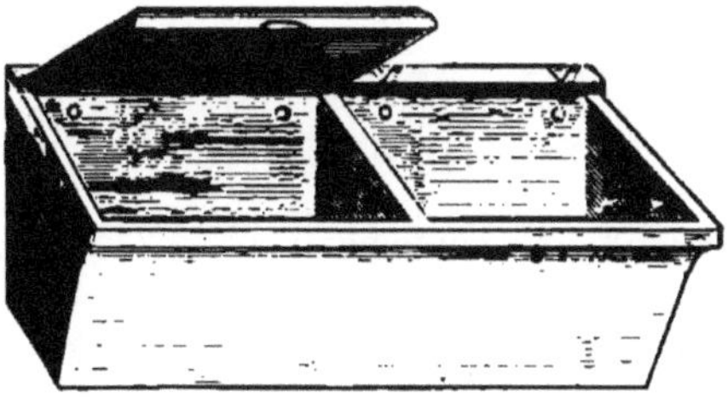

Single Tub.

NUMBER.	LENGTH.	WIDTH.	DEPTH.	PRICE.
1	25 in.	21 in.	16 in.	$ 8.00
2	25 "	24 "	16 "	8.50
3	27 "	24 "	16 "	9.00
4	31 "	24 "	16 "	11.00

Two Part Tub.

11	48 in.	21 in.	16 in.	$12.50
12	48 "	24 "	16 "	15.50
13	53 "	24 "	16 "	16.50
14	60 "	24 "	16 "	17.50

Three Part Tub.

21	72 in.	21 in.	16 in.	$22.00
22	72 "	24 "	16 "	26.50
23	80 "	24 "	16 "	27.00
24	90 "	24 "	16 "	29.00

State if Right or Left Waste wanted for above Tubs.

With Patent Metalic Casing and Patent Waste Connection.

The connection on this Tub can be made from $4 to $6 cheaper than on any other Tub manufactured, as plugs and connections are cast on Tub.

Single Tub.

NUMBER.	LENGTH.	WIDTH.	DEPTH.	PRICE.
51	25 in.	21 in.	16 in.	$ 8.75
52	25 "	24 "	16 "	9.25
53	27 "	24 "	16 "	9.75
54	31 "	24 "	16 "	11.75

Two Part Tub.

61	48 in.	21 in.	16 in.	$14.25
62	48 "	24 "	16 "	17.25
63	53 "	24 "	16 "	18.25
64	60 "	24 "	16 "	19.25

Three Part Tub.

71	72 in.	21 in.	16 in.	$24.75
72	72 "	24 "	16 "	29.25
73	80 "	24 "	16 "	29.95
74	90 "	24 "	16 "	31.75

These Tubs can be used either right or left.

Ash Covers.

1 Part....................$1.50	2 Part....................$3.00	3 Part....................$4.50

Soap Stone Wash Trays

Outside Measure.	Length.	Width.	Depth.	Price without Legs.
2 part..............................	4 feet.	2 feet.	16 inches.	$22.00
2 "	4½ "	"	"	26.00
2 "	5 "	"	"	28.00
2 "	5½ "	"	"	30.00
3 "	6 "	"	"	34.00
3 "	6½ "	"	"	40.50
3 "	7 "	"	"	42.00
3 "	7½ "	"	"	45.00
3 "	8 "	"	"	51.00

Soap Stone Sinks.

Length.	Width.	Depth.	Price.
2 feet.	18 inches.	8 inches.	$ 7.00
2½ "	19 "	"	9.00
3 "	20 "	"	10.00
3½ "	22 "	"	12.00
4 "	24 "	"	15.00
4½ "	24 "	"	17.00
5 "	24 "	"	18.50

Iron Legs for Soap Stone Wash Trays.

Plain, each..............................$1.40 | Galvanized, each......................$2.75

Iron Brackets for Soap Stone Sinks.

Plain, each..............................$1.40 Galvanized, each......................$2.75

White Crockery Stationary Wash Tubs.

Measurement inside from right to left 25 to 26 inches; front to back on top 21½ inches; depth 14½ inches; 1¼ inch thick with a heavy 1¾ inch flange all around.

1 Tub, without Stand or Top	$20.00
2 Tubs, with Galvanized Iron Stands and Hard Wood Tops	51.00
3 " " " " " " " " "	75.00

White Crockery Sinks.

Made of same material, strength and durability as the Tubs.

24x18x 8,	Butler's Pantry, each		$ 8.00
30x20x 7,	Kitchen,	"	12.00
37x22x 7½,	"	"	16.00
19x19x15,	Slop,	"	14.00

PASTE OVER PAGE 134.

REVISED STANDARD LIST OF AMERICAN SANITARY EARTHENWARE.

Adopted August 22nd, 1895.

	12	13	14	15	16 in.
Round Basins, no Overflow	.55	.70	.75	$1.20	$1.55
" " Common Overflow	.70	.83	.90	1.35	1.85
" " Patent "	.77	.90	1.00	1.50	2.00
" " " " for Rubb'r Plug	.95	1.05	1.15	1.70	2.20

	14x17	15x19	16x21 inch.
Oval Basin, no Overflow	$2.10	$3.35	$4.10
" " Common Overflow	2.10	3.35	4.10
" " Patent "	2.50	3.50	4.25
" " " " for Rubber Plug	2.80	3.80	4.55
Add for Embossing	.75	.75	.75

Round French Closet Bowl	$ 1.35
Oval Closet Bowl	1.50
Round Pipe Wash Closet Bowl	1.50
" " " and Vent Closet Bowl	1.65
Philadelphia Hopper	2.65
Ship Closet Basin, No. 1	1.50
" " " No. 2	1.40
" Plug Basin, 13 inch	1.05
" " " 14 inch	1.20
Bidet Pan, for Brass Plug	2.90
" " " " " with Overflow	3.00
" " " " " no Hole	2.65
Bedfordshire Urinal, No. 1, Flat	5.85
" " " 2, "	4.40
" " " 3, "	3.95
" " " 1, Corner	5.85
" " " 2, "	4.40
" " " 3, "	3.95
" " " 1, Flat with Lip	8.80
" " " 2, " " "	7.30
" " " 3, " " "	6.60
" " " 1, Corner, with Lip	8.80
" " " 2, " " "	7.30
" " " 3, " " "	6.60
No. 1, F. B. Lipped Vent and Hood Urinal	10.25
" 2, " " " " " "	8.80
" 1, Corner " " " " "	10.25
" 2, " " " " " "	8.80
Oval Short Hopper and Trap Combined	8.05
" " " " " Embossed	8.80
Square " " " " Combined	9.50
" " " " " Embossed	10.25
Oval Tall F. R. Hoppers	5.55
" " " " with Seat Vent	5.70
" " " " " Hub "	6.00
Round Tall F. R. Hoppers	3.70
" " " " with Seat Vent	3.85
" " " " " Hub "	4.15

PASTE OVER PAGE 135.

Square Tall F. R. Hoppers, 14x14	6.95
" " " " with Seat Vent	7.10
" " " " " Hub "	7.40
Square Long Hopper, 16x16	11.10
Oval Short F. R. Hoppers	2.65
" " " " with Seat Vent	2.80
Round Short F. R. Hoppers	1.90
" " " with Seat Vent	2.05
Square Short F. R. Hoppers	4.40
" " " " with Seat Vent	4.55
Oval F. R. Closet Bowl	2.05
" " " " with Seat Vent	2.20
Traps	2.45
Drip Trays, Oval and Round	2.60
Hopper Stands	1.30
" " with Hub Vent	1.60
Floats, Large, above 7 inch	.44
" Small, under 7 inch	.38

Closets and Pedestals.

WASHOUTS PLAIN.

No. 1, Oval Front Outlet	$13.20
" " Back "	13.90
" 2, " Front "	10.25
" " Back "	11.00
" 3, " Front or Back Outlet	8.05
Square Front Outlet	15.35
" Back "	16.10
No. 1, Square-Back, Front Outlet	14.05
" " " Back "	14.80
" 2, " " Front "	12.55
" " " Back "	13.30
" 1, Washouts for Iron Traps	7.20
" 2, " " " "	6.60

WASHOUTS EMBOSSED.

No. 1, Oval Front Outlet	$13.90
" " Back "	14.65
" 2, " Front "	11.00
" " Back "	11.70
" 3, " Front or Back Outlet	8.80
Square Front Outlet	16.10
" Back "	16.85
No. 1, Square-Back, Front Outlet	14.80
" " " Back "	15.50
" 2, " " Front "	13.30
" " " Back "	14.05

PEDESTALS PLAIN.

No. 1, Oval, Front Outlet	$13.90
" 2, " " "	11.00
" 3, " " "	8.80
Square, Front Outlet	16.10
" Back "	16.85
No. 1, Square-Back, Front Outlet	14.80
" 2, " " " "	13.30
" 1, " " Back "	15.50
" 2, " " " "	14.05

PEDESTALS EMBOSSED.

No. 1, Oval, Front Outlet	$14.65
" 2, " " "	11.70
" 3, " " "	9.50
Square, Front Outlet	16.85
" Back "	17.55
No. 1, Square-back, Front Outlet	15.50
" 2, " " " "	14.05
" 1, " " Back "	16.20
" 2, " " " "	14.75

	14x17 Price	15x19 inch. Price
Square Basins, no Overflow	$4.70	$5.70
" " Common Overflow	4.70	5.70
" " Patent Overflow	5.00	6.05
" " " " for Rubber Plug	5.35	6.35
Add for Embossing	.75	.75

	14x17	15x18 inch.
	Price.	Price.
Square Basins, no Overflow	7.00	8.50
" " Common Overflow	7.00	8.50
" " Patent Overflow	7.50	9.00
" " " " for Rubber Plug	7.95	9 45
Recess Basins, Oval	7.00	8.50

	14	15	16 in.
	Price.	Price.	Price.
Recess Basins, Round	5.00	6.00	7.00

	14x17	15x19 inch.
	Price.	Price.
Recess Basins, Embossed	8.00	9.50

	14	15	16 in.
	Price.	Price.	Price.
Recess Basins, Round, Embossed	6.00	7.00	8.00

Metal Connections, Net.

SIZE	Plain.	Polished.
1 inch, Straight	.30	.50
1 " Bent	.35	.55
1¼ " Straight	.35	.55
1¼ " Bent	.40	.60
1½ " Straight	.45	.65
1½ " Bent	.50	.70
2 " Straight	.60	.80
2 " Bent	.65	.85
Small Clamp	.07	.....
Hand-Hole Clamp and Washer	.25	.....

Packages to be Charged as Follows, Net:

Crates	$2.00
Hhds., Extra Large	2.00
" Large	1.50
" Medium	1.25
" Small	1.00
Tierces, Large	1.00
" Small	.75
Boxes	.50
Barrels	.35

WATER CLOSET APPARATUS.

We furnish to order any of the Closets made by the leading manufacturers **at their prices.** We carry in stock many of the leading kinds, of which the following is a partial list only:

Philadelphia Iron Hoppers.

With Side Arm.

Plain	$1.10
Enamelled	1.50
Enamelled, with Flushing Rim	5.00

Iron Hopper Valve Closet.

With Self-Raising Round Seat, Single or Double-Acting Valve.

Works under any pressure from 5 to 200 lbs. No reaction. The valve stem can be turned to regulate the flow of water.

Porcelain lined	$8.00

Iron Hopper Valve Closet.

With Self-Raising Round Seat, Single or Double-Acting Valve.

Works under any pressure from 5 to 200 pounds. Valve stem can be turned to regulate flow of water.

Porcelain Lined Hopper, Plain Trap	$ 9.60
" " " and Trap	10.00

Can also be furnished on ¾ or ½ S. Trap, if desired.

Valve Pan Closet.

Adapted to any pressure, by a regulating screw in the valve.

Price, with Brass Cup and Pull	$4.65
" " Nickel Plated "	5.00

Plunger Valve Closet.

Safety	$13.00
Advance	18.00
Detroit Sanitary	16 00

For Porcelain Lined Valve Section, add $4 00.
No extra charge for Vented Bowl.

Tall Earthen Hoppers.

Phila. Large Flange	$ 3.75
" Flushing Rim	7.25
Square Top, " "	13.15
" " " " with Seat Vent.	13.40
Oval " " "	10.50

Couplings Extra.

Short Oval Earthen Hoppers.

Oval Flushing Rim	$5.00
Vented	5.30

Couplings Extra

Traps for Short Oval Hoppers.

Earthen	$4.60
Iron, Plain	1.50
" Enameled	2.00

Couplings Extra.

Short Oval Earthenware Hopper.

With Earthen Trap and Flushing Rim.

Oval Bowl	$9.60
" " Vented	9.90

Couplings Extra.

Plain, Oval, Front-Outlet Washout Closet.

Three Sizes.

No. 1	$22.35
" 2	21.05
" 3	18.40

Couplings Extra.

Plain, Oval, Back-Outlet Washout Closet.

Three Sizes.

No. 1	$22.35
" 2	21.05
" 3	18.40

Couplings extra.

"Creole."

No. 3 Front-Outlet Embossed Washout Closet.

Price	$21.05

Couplings Extra.

"Hero."

WASHOUT CLOSET.

No. 2 Front-Outlet, Pedestal Embossed.

Price	$26.30

Couplings Extra.

Syphon Jet Water Closet.

Pronounced by leading experts to be in compliance with natural laws and the best sanitary fixture of the age. Full contents of closet ejected instantly direct to the sewer. Has deep seal in bowl against sewer gas. Requires no trap.

Price, Embossed	$35.00
" Plain	32.50

Couplings Extra.

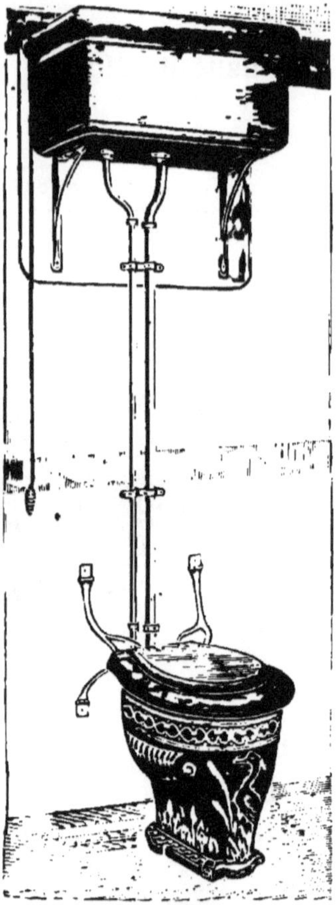

Stork Closet.

This cut shows our Stork Closet with a handsome design of Tank, Seat and Nickel Plated Flush and Supply Pipe.

Price, complete	$60.00
Add, for Floor Slab	10.00

Rubber Elbows for Closets.

The Closet can be connected to Flush Pipe from Tank by simply Expanding Elbow over the Pipe: it makes a secure Joint, and after rubber is set to Pipe it can't be pulled off.

No. 2, $12.00 Dozen.

No. 3, $9.00 Dozen.

Adjustable Couplings.

If too long, cut off in line of rear flange, and slip nut over.

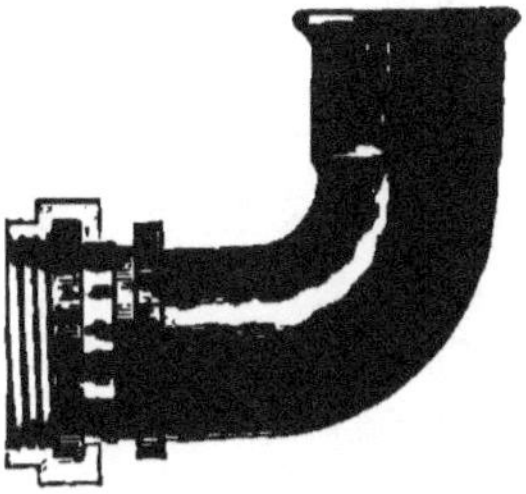

No. 4, $11.00 Dozen.

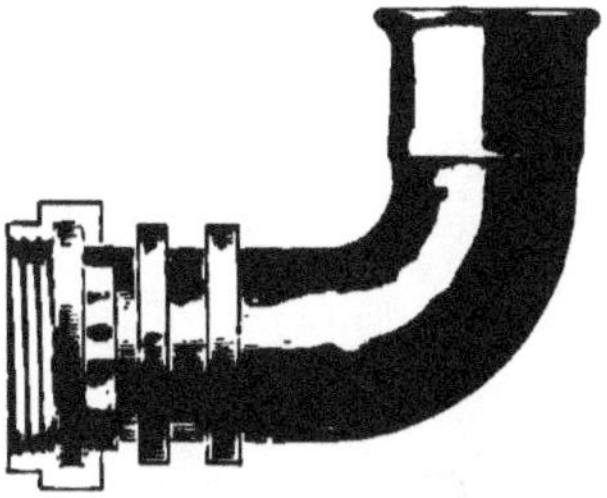

No. 4 A, $12.00 Dozen.

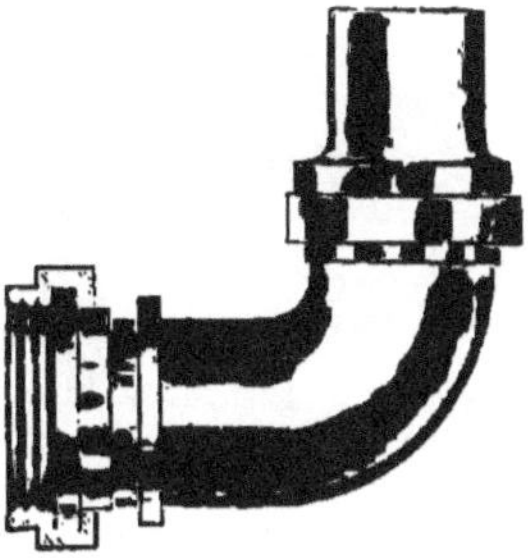

No. 5, $14.00 Dozen.

No. 8, for 2 Inch Vent, $14.00.

No. 6	Elbow, for 2 inch Vent, same shape as No. 3 or 4, per dozen	......	$15.00
" 6A	" 2½ " " " " " "	....	15.00

Drip Trays.

Round or Oval, Sheet Iron, Enamelled Both Sides.

Price, Each $1.50

N. B.—In ordering state whether Round or Oval is wanted.

CLOSET TANKS.

Plain Iron with Cage Valve, for Hoppers and Low Priced Closets.

No. 1. Small..$4.00
" 2, Large.. 6.00

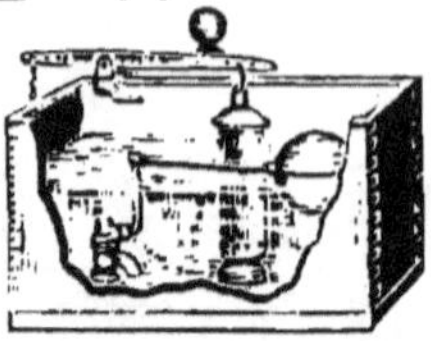

"Syphon"

Copper-Lined.

Plain..$7.50

Gold Medal.

Can Be Regulated to Give as Long or Short Flush as Desired. Copper Lined.

Plain..$8.00

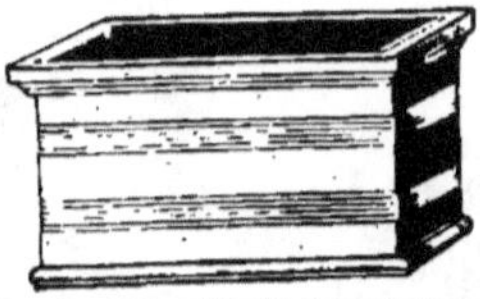

Beaded..$9.00

This cut shows our Cabinet Finish Beaded Tank, with Gold Medal or Syphon Valve, finished in Cherry, Antique, or Natural Oak, Black Walnut or Ash, and makes a very neat finish at a low price.

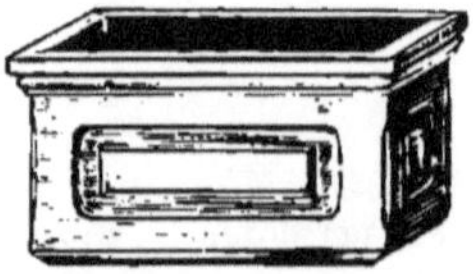

Embedded Panel..$10.00

This cut shows our Embedded Panel Tanks with Syphon Valve, finished in Cherry, Antique, Natural Oak, or Black Walnut. Packed one in a box.

Framed Panel..$12.00

This cut represents one of our best cabinet finished Tanks with Syphon Valve, and is suitable to be put up where parties desire a first-class Tank. We can furnish it finished in Cherry, Natural and Antique Oak, or Black Walnut.

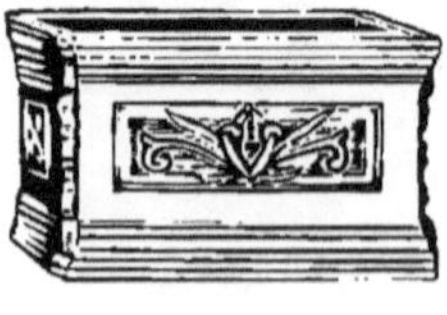

Hand Carved Tank..$13.50

With Gold Medal Valve that can be regulated to give a long or short flush, as desired.

They are finished in Cherry, Natural or Antique Oak, and Black Walnut, and make a first class Tank in finish. Packed one in a box.

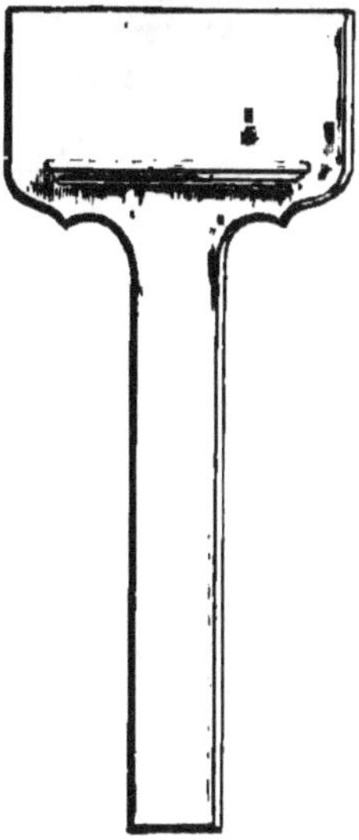

Pipe Boards.

Ash; Dark or Light Cherry; Plain Oak; Finished Antique or Natural..........................$3.75
Cherry; Walnut; Quartered Oak; Finished Antique or Natural.. 5 00

No Brackets required. Each Board is provided with two Castings for fastening Tank to the Board.

Flush and Supply Pipes.

	Nickel Plated.	Brass Finished.
1¼ inch Flush Pipe, 6 feet long.......	5.00	4.75
1½ " " 6 " "	6.50	6.25
⅝ " Supply Pipe, 7 " "	3.75	3 50

Closet Seats.

For Oval and Round Pedestal Closets.

This cut represents our No. 3 Seat, 1 inch thick, with Beaded Back, finished in Natural Cherry, Natural Oak, Antique Oak, Ash, Dark Cherry, and Black Walnut.

No. 1 Seat	$2.50
" 2 " with Back	3.50
" 3 " " " and Cover	5.50

This cut represents our No. 3 Seat, 1¼ inch thick, finished in Natural Cherry, Natural Oak, Antique Oak, Dark Cherry, Black Walnut and Ash.

No. 2 Seat, no Cover	$4.50
" 3 " with Cover	6.50

To parties desiring a first-class Seat, both in style and finish, we advise this; it being built in a very substantial manner, and at the same time being very showy, the back is extra high, and framed together, forming a solid panel.

The foregoing Tanks and Seats comprise only a small line of Samples of the various styles we keep in stock. We also furnish to order any of the styles made by all manufacturers at **their prices.**

Closet Valves.

Valve for Pan Closet	$2.25
" " Safety Plunger Closet	3.50
Moore's Small Cage Valve, for Closet Tank	1.00
" Time " " "	3.50
Dalton & Ingersoll Time Valve, for Closet Tank	4.00
" " Cistern " " "	1.75
Gold Medal Valves, for Closet Tank	2.00
Syphon " " "	2.00
Clamps for Earthen Hoppers Traps	.10

Flush and Supply Pipe Straps.

	Nickel Plated.	Brass Finished.
1¼ inch Double Straps for Flush and Supply Pipes, per set.....	.90	.80
1½ " " " " " " "	1.00	.90
1¼ " Single Straps for Flush Pipe only, per set.................	.75	.65
1½ " " " " " " "	.85	.75

Brackets.

Designed for Closet Tanks and Seats, and Marble Slabs.

No. 1.

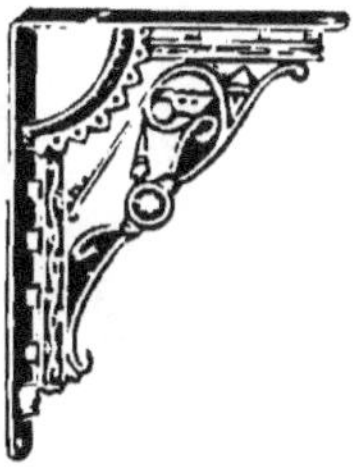

No. 2.

No. 3.

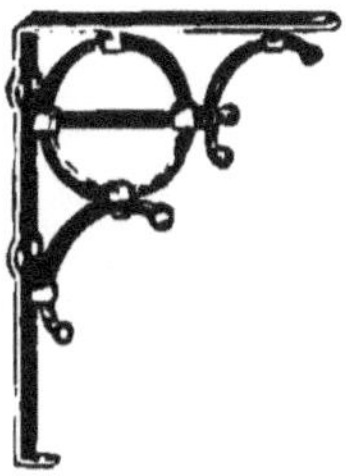

No. 4.

SIZE................................	5x7	7x9	9x12 inch.
No. 1, Iron, English Maroon, per dozen pair	1.25	1.50	1.75

SIZE................................	5x7	7x9	9x11	16x18 inch.
No 2, Iron Brass Plated, per dozen pair..................	4.10	6.00	8.25	32.00
" 2, " Nickel " " "	4.10	6.00	8.25	32.00

SIZE................................	4x5	7x9	9x11	16x18 inch.
No. 3, Brass, Polished, per dozen pair......................	10.75	17.50	21.50	46.75
" 3, " Nickel Plated, per dozen pair..............	12.00	20.00	24.00	51.00

SIZE................................	16x18	20x28 inch.
No. 4, Polished Brass, per dozen pair..............................	70.00	135.00
" 4, Nickel Plated..	74.00	140.00

The 16x18 Bracket is same as cut, except has no foot.

Iron Wash Stand.—On Frame.

Patent Overflow and Rubber Plug.

Plain	$ 5.00
Painted	5.50
Galvanized	7.75
Enameled Slab with Bronzed Frame	8.50
Frame	2.00

Double Wash Stand.—On Frame.

Patent Overflow and Rubber Plugs.

Similar to above.

Plain	$ 8.00	Enameled Slab and Bronzed Frame	$17.00
Painted	9.00	Frame	3.00
Galvanized	14.00		

Iron Corner Wash Stand.

ON STANDARD.

Patent Overflow and Rubber Plug.

Plain	$4.25
Painted	4.75
Galvanized	6.75
Enameled Slab with Bronzed Standard	7.50

Iron Corner Slab and Bowl.

With Patent Overflow and Rubber Plug.

Plain	$2.25
Painted	2.75
Galvanized	3.75
Enameled	5.50

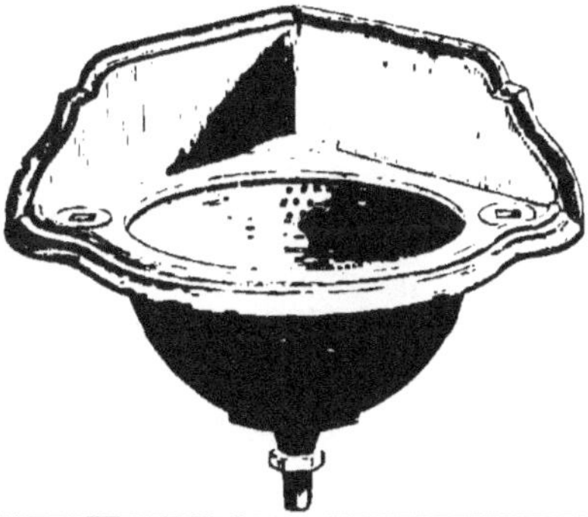

Iron Half-Circle Wash Stand.

ON STANDARD.

Patent Overflow and Rubber Plug.

Plain	$4.25
Painted	4.75
Galvanized	6.75
Enameled Slab with Bronzed Standard	7.50

Half-Circle Iron Slab and Bowl.

With Patent Overflow and Rubber Plug.

Plain	$2.25
Painted	2.75
Galvanized	3.75
Enameled	5.50

Patent Overflow Iron Wash Basins.

SIZE	$10\frac{1}{2}$	$11\frac{3}{4}$	14 in.
Plain	1.25	1.50	1.75
Painted	1.50	1.65	1.90
Galvanized	2.00	2.25	2.50
Enameled	2.50	2.75	3.00

Common Overflow Iron Wash Basins.

SIZE	10½	11¾	14 in.
Plain	1.00	1.20	1.40
Painited	1.15	1.35	1.55
Galvanized	1.75	2.00	2.25
Enameled	2.25	2.50	2.75

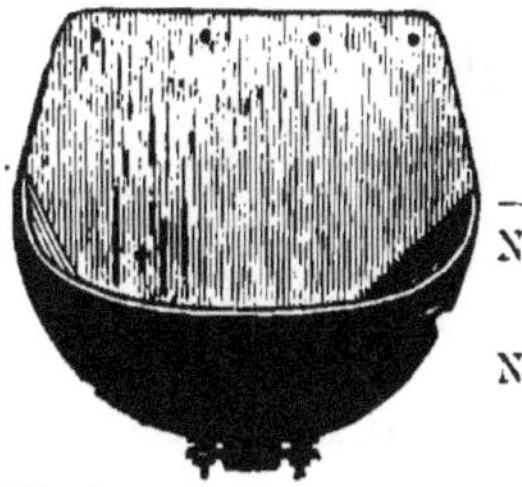

Half-Circle Urinals.

No. 1.—12 in. on Back,	Plain	$	.75
" "	Galvanized		1.50
" "	Enameled		1.90
No. 2.—15 " "	Plain		1.00
" "	Galvanized		1.90
" "	Enamelled		2.25

Corner Urinals.

No. 1.—7 in. on Side,	Plain	$	.60
" "	Galvanized		1.00
" "	Enameled		1.40
No. 2.— " "	Plain		.75
" "	Galvanized		1.25
" "	Enameled		1.70
No. 3.—12 in. "	Plain		.90
" "	Galvanized		1.65
" "	Enameled		2.10

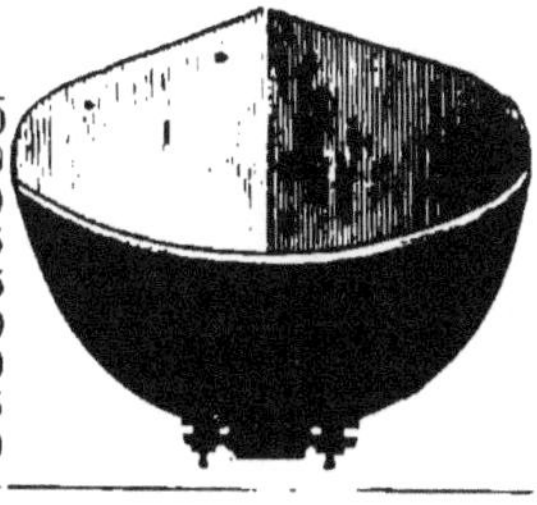

PLUMBERS' IRON WORK.

CAST IRON SINKS.

We also furnish Sinks with Patent Overflow when required

	Old List.			New List.		
Size.	Painted.	Galvanized	Enameled	Size.	Depth.	Painted
14x14, 6 inches deep, each,	1.90	....	....	12x18 in.	6 in	1.70
16x16, 6 " " "	2.25	3 25	5.25	12x20 "	6 "	1.85
18x12, 6 " " "	1.65	2.00	3 90	13x19 "	5 "	1.75
19x13, 4½ " " "	1.75	......	......	14x20 "	6 "	2.20
20x14, 5 " " "	2.20	......	......	14x22 "	6 "	2.25
22x14, 6 " " "	2.30	2.60	4.60	14x24 "	6 "	2.30
23x15, 3½ " " "	2.30	2.70	5 00	15x23 "	6 "	2.40
25x14, 5 " " "	2.40	......	......	16x24 "	6 "	2.70
24x16, 6 " " "	2.70	3.00	5 25	15x25 "	6 "	2.75
24x18, 5 " " "	2.70	......	......	15x27 "	6 "	2.90
24x18, 6 " " "	2.80	3.50	5.50	16x28 "	6 "	3.20
27x15, 5 " " "	2.90	3.50	5.50	16x30 "	6 "	3.50
28x16, 6 " " "	3.20	4.00	5.75	17x25 "	6 "	3.20
30x17, 4 " " "	3.00	......	......	17x28 "	6 "	3.35
30x16, 6 " " "	3.50	3.60	6.00	18x24 "	6 "	2.80
30x17, 5 " " "	3.60	......	......	18x30 "	6 "	3.75
30x18, 6 " " "	3.75	4.00	6.50	20x30 "	6 "	4.00
30x20, 5 " " "	3.75	......	......	18x32 "	6 "	4.25
30x20, 6 " " "	4.00	4.75	7.00	18x36 "	6 "	4.75
35x19, 4½ " " "	4.00	......	......	19x38 "	6 "	5.00
32x18, 6 " " "	4.25	4.75	7.25	20x36 "	6 "	5.25

CAST IRON SINKS.—Continued.

Size.	Old List. Painted.	Old List. Galvanized.	Old List. Enameled	New List. Size.	New List. Depth	New List. Painted
35x17, 6 inches deep, each,	4.50	......	......	20x40 in.	6 in.	6.00
36x18, 6 " " "	4.75	5.00	7.25	18x42 "	6 "	6.25
36x22, 5 " " "	5.40	......	......	20x42 "	6 "	6.50
36x20, 6 " " "	5.25	......	......	22x42 "	6 "	6.75
36x22, 6 " " "	5.50	6.25	8.00	24x48 "	6 "	8.00
38x19, 6 " " "	5.60	6.50	8.50	24x50 "	6 "	8.50
40x20, 6 " " "	6.00	......	......	....	......	......
42x21, 4½ " " "	6.00	......	......	....	......	......
42x18, 6 " " "	6.25	......	......	....	......	......
42x23, 5 " " "	6.25	......	......	....	......	......
42x23, 6 " " "	6.50	7.00	9.50	....	......	......
44x22, 6 " " "	6.75	......	......	....	......	......
48x22, 5 " " "	7.00	......	......	....	......	......
48x23, 6 " " "	7.50	9.50	12.00	....	......	......
50x20, 5 " " "	8.00	......	......	....	......	......
50x24, 6½ " " "	8.50	14.00	16.00	....	......	......
52x26, 6½ " " "	10.00	......	......	....	......	......
56x32, 9 " " "	16.00	29.00	29.00	....	......	......
62x22, 8 " " "	10.75	18.50	23.00	....	......	......
76x22, 7 " " "	15.00	26.00	28.00	....	......	......
60x28, 10 " " "	18.00	34.00	34.00	....	......	......
78x28, 10 " " "	25.00	45.00	45.00	....	......	......
94x24, 10 " " "	30.00	60.00	......	....	......	......

Patent Extension Sinks.

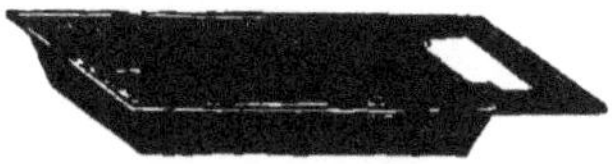

PAINTED.

With Extension to Set Pump on.

No. 5, 18x32, 5 inches deep, each	$3.50
" 6, 20x38, 5 " " "	4.40
" 7, 22x44, 5 " " "	5.50
" 8, 23x50, 5 " " "	6.50

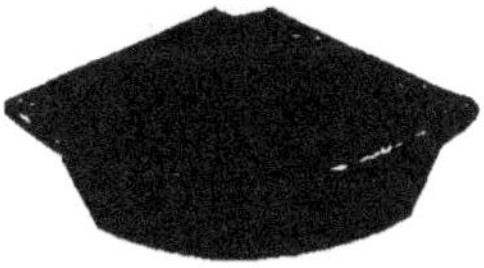

Corner Sinks.

	Painted.	Galvanized.	Enameled.
No. 1, side 20 in., front 29 in., depth 6 in., each...	2.50	2.20	5.00
" 2, " 22 " " 31 " " 6 " " ...	3.15	2.75	5.75

Half-Circle Sinks.

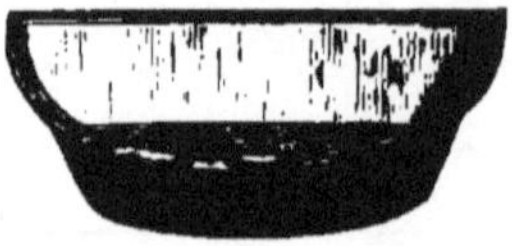

	Painted.	Galvanized.	Enameled.
No. 1, back 24 in., width 14 in., depth 6 in., each,	2.50	2.60	5.00
" 2, " 27 " " 15 " " 6 " "	3.15	3.00	6.00
" 3, " 31 " " 17 " " 6 " "	4.00	4.25	7.00

Slop Sinks, With or Without Bell Trap.

The Galvanized and Enameled are Made Only Without Bell Traps.

	Painted.	Galvanized.	Enameled.
No. 1, 16x16, 10 inches deep, each........	4.00	4.25	5.75
" 2, 14x20, 12 " " "	5.00	5.50	7.25
" 3, 20x24, 12 " " "	6.00	7.75	10.00
" 4, 20x30, 12 " " "	11.00	12.50	14.50

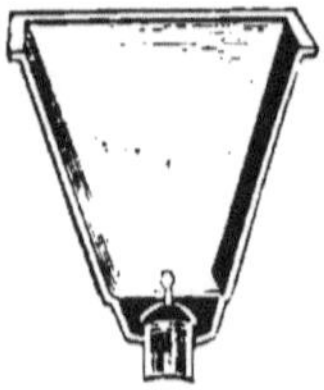

Improved Sewer Trap and Slop Sink.

With Trap and Strainer.

No. 1, 12x12 in., 6 in. deep, 2 in. outlet, each..............	$2.25
" 2, 15x15 " 11½ " " 2 " " "	3.35
" 3, 18x18 " 12 " " 3 " " "	4.25
" 4, 20x20 " 12 " " 3 " " "	5.25

Cellar Traps.

No. 1, 9x9 in., 2¼ in. deep, 2 in. outlet................	$1.25
" 2, 12x12 " 2¼ " " 2 " "	1.75

Sewer Traps.

No. 1, 16x16, 10 inches deep, each	$2.50
" 2, 20x20, 12 " " "	3.50
" 3, 20x20, 15 " " "	4.50

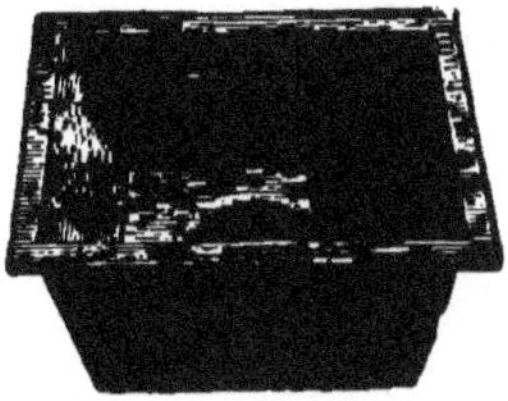

Hydrant Cess-Pools.

For Use in Front of Pumps and Hydrants.

No. 1, 14x14, 6 in. deep, with Bell Trap, each	$2.00
" 2, 16x16, 6 " " " "	2.25
12x12, 6 " without " "	1.00
14x14, 6 " " " "	1.15
16x16, 6 " " " "	1.30

Stable Cess-Pool.

With Bell Trap.

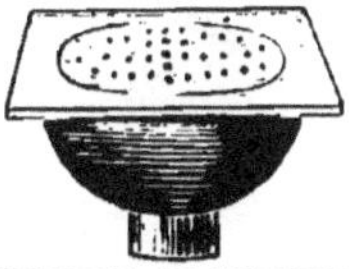

Size 6x6, painted each	$ .75
" 9x9, " "	1.00
" 13x13, " "	3.00

Sink Back Without Air Chambers.

	Plain	Galvanized.	Enameled		Plain.	Galvanized.	Enameled.
Length, 20 in.	1.00	1.75	2.75	Length 36 in. ...	2.25	4.25	5.50
" 22 "	1.15	2.00	3.10	" 38 " ...	2.50	4.50	6.00
" 23 "	1.20	2.10	3.20	" 42 " ...	2.75	4.75	6.50
" 24 "	1.25	2.25	3.25	" 48 " ...	3.00	5.50	7.50
" 25½ "	1.30	2.40	3.75	" 50 " ...	4.00	6.50	8.50
" 27 "	1.35	2.60	4.00	" 62 " ...	5.00	8.50	10.00
" 28 "	1.40	2.90	4.25	" 76 " ...	6.00	11.00	13.00
" 30 "	1.50	3.00	4.50	" 94 " ...	8.00	14.00	16.00
" 32½ "	1.75	3.25	5.00	" 120 " ...	11.00	18.00	22.00

Sink Back With Air Chamber, Pipes and Couplings.

To prices of Backs add for each air chamber:

Plain	$1.50
Galvanized	2.00
Enameled	2.50

Cellar Drainer.

With Automatic Movements.

No. 1,	Automatic Movement......	$ 25.00	Without Automatic Movement..	$ 15.00
" 2,	" "	40.00	" " "	25.00
" 3,	" "	55.00	" " "	35.00
" 4,	" "	80.00	" " "	50.00
" 5,	" "	110.00	" " "	70.00
" 6,	" "	160.00	" " "	100.00

Circular furnished on application.

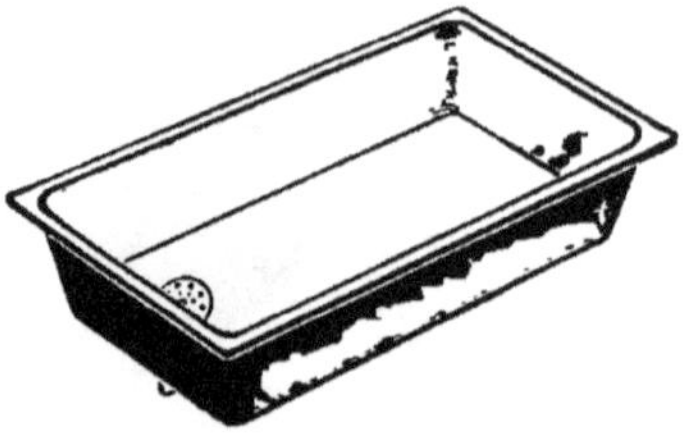

Steel Sinks.

	Painted.	Galvanized	White Enamel.	Gray Enamel.
Size, 16x24, each........	1.80	4.00	7.50	6.50
" 18x30, "	2.50	5.10	10 00	8.50
" 18x36, "	3.00	6.50	11.00	9 50
" 20x30, "	3.00	6.25	10.50	9.00
" 20x36, "	3.70	7.75	12.00	10.50
" 20x40. "	4.00	8.50	13.00	11.50
" 14x20, Ovaled Sink, each........	2.00	3.50	5.50	6.50

COUPLINGS FOR STEEL SINKS.

Brass Couplings, with Strainer, for both lead and wrought iron pipe, each, *net*, $ .50
" " " Powell's Patent Improved Star Metal Cored Rubber Stopper, for both lead and wrought iron pipe, each, net............ .75
Oval Sinks, with Patent Overflow, extra, *net*, each............ .50

Sink Legs.

Plain, price, each............ $ .38
Galvanized, " " .... .75

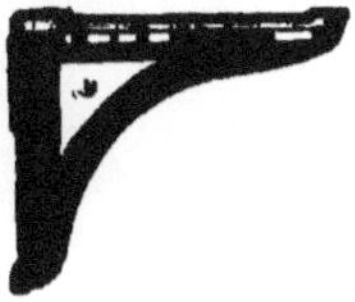

Sink Brackets.

	Plain.	Galvanized.
No. 1, Takes Sink up to 18 in. wide, each......	.38	.75
" 2, " " " 23 " "	.50	1.00
" 3, " " " 32 " "	1.00	1.25

Fixtures for Cast Iron Sinks.

	Plain.	Galvanized.	Enameled.
Sink Bolts...................per 100,	1.50	......	......
Sink Strainers...................per dozen,	1.25	2.25	2.50
" " with Plug.................. "	2.75	4.25	5.00
Sink Couplings.................. "	1.25	1.75	.....

Iron Traps for Sinks.

Half S, Three-quarter S, or full S, for Lead Pipe Connection, each...........	$1.25
" " " " " Iron " " "	1.25
Straight Sink Connections, " " " " "	1.00

Boiler Stand.

This Standard can be lowered at pleasure by cutting off the ring on Standard, as shown by the cut.

Size..........................	Plain.	Galvanized.
12 inch Ring, each...........	1.25	2.50
13 " " "	1.30	2.60
14 " " "	1.40	2.70
15 " " "	1.50	3.00
16 " " "	1.75	3.25
18 " " "	2.00	3.60
20 " " "	2.25	4.25
22 " " "	2.50	4.75
24 " " "	3.00	5.50

PUMPS.

Pitcher Pumps.

FITTED FOR EITHER LEAD OR IRON PIPE.

Pitcher Spout.—Open Top.

No. 1, 2½ in.,	suitable for pipe	1¼ or 1½	in. calibre	$4.25
" 2, 3 "	" "	1¼ or 1½	" "	4.75
" 3, 3½ "	" "	1¼ or 1½	" "	5.25
" 4, 4 "	" "	1¼ or 1½	" "	5.75

Pitcher Spout.—Closed Top.

No. 1, 2½ in.,	suitable for pipe	1¼ or 1½	in. calibre	$4.25
" 2, 3 "	" "	1¼ or 1½	" "	4.75
" 3, 3½ "	" "	1¼ or 1½	" "	5.25
" 4, 4 "	" "	1¼ or 1½	" "	5.75

Closed Spout.—Open Top.

No. 1, 2½ in.,	suitable for pipe	1¼ or 1½	in. calibre	$4.25
" 2, 3 "	" "	1¼ or 1½	" "	4.75
" 3, 3½ "	" "	1¼ or 1½	" "	5.25
" 4, 4 "	" "	1¼ or 1½	" "	5.75

Closed Spout.—Closed Top.

No. 1, 2½ in.,	suitable for pipe	1¼ or 1½	in. calibre	$4.25
" 2, 3 "	" "	1¼ or 1½	" "	4.75
" 3, 3½ "	" "	1¼ or 1½	" "	5.25
" 4, 4 "	" "	1¼ or 1½	" "	5.75

Cistern Pumps—On Base.

FITTED FOR EITHER LEAD OR IRON PIPE.

Iron.

No. 0, 2 in.,	suitable for pipe	1	in. calibre	$3.50
" 1, 2¼ "	" "	1¼	"	4.00
" 2, 2½ "	" "	1¼	"	4.50
" 3, 2¾ "	" "	1¼	"	5.00
" 4, 3 "	" "	1¼	"	5.50
" 5, 3¼ "	" "	1½	"	6.50
" 6, 3½ "	" "	2	"	8.00

Brass Cylinder.

No. 1, 2¼ in.	diameter, suitable for	1	in. pipe	$6.00
" 2, 2½ "	" " "	1¼	" "	7.00
" 3, 2¾ "	" " "	1¼	" "	8.00
" 4, 3 "	" " "	1¼	" "	10.00
" 5, 3¼ "	" " "	1½	" "	14.00
" 6, 3½ "	" " "	2	" "	18.00

Anti-Freezing Shallow Well Pumps.

OPEN TOP.

Screwed Cylinder.

No. 2, 2½	in. calibre,	for pipe	1¼	in. calibre,	each	$ 7.75
" 3, 2¾	"	"	1¼	"	"	8.00
" 4, 3	"	"	1¼	"	"	8.50
" 5, 3¼	"	"	1¼	"	"	9.00
" 6, 3½	"	"	1¼	"	"	10.00

Bolted Cylinder.

No. 2, 2½	in. calibre,	for pipe	1¼	in. calibre,	each	$ 8.00
" 3, 2¾	"	"	1¼	"	"	8.25
" 4, 3	"	"	1¼	"	"	8.50
" 5, 3¼	"	"	1¼	"	"	9.25
" 6, 3½	"	"	1¼	"	"	10.25

Anti-Freezing Force Pump.

The Piston and Valve are located three feet below the base, at which point there is a vent-hole for allowing the water to pass off below freezing point. Fitted for Wrought Iron Pipe, 1¼ inches calibre, unless otherwise ordered. We furnish an attachment for connecting Hose to Spout, with each pump.

Price, complete, with 3 inch Cylinder	$13.50
" " " 3½ " "	14.50
Hose and Discharge Pipe, extra, net	2.00

Force Pumps—Double Discharge—On Base.

Iron.

No. 1, 2½	in. diameter, suitable for	1	in. pipe	$11.00
" 2, 2¾	" " "	1 and 1¼	"	12.00
" 3, 3	" " "	1¼	"	13.00
" 4, 4	" " "	1½ and 2	"	20.00

Add, for Iron Cock on Side Outlet, $2.50.

Brass Cylinder.

No. 1, 2½	in. diameter, suitable for	1	in. pipe	$16.00
" 2, 2¾	" " "	1 and 1¼	"	16.50
" 3, 3	" " "	1¼	"	17.00
" 4, 4	" " "	1½ and 2	"	32.00

The above Pumps are also made with Brackets to screw to wall or on a board, at same price.

Force Pumps.—With Spouts.—On Base.

Iron.

No. 1, 2½	in. diam., suitable for	1		in. pipe,	$10.00
" 2, 2¾	" "	" 1	and 1¼	"	11.50
" 3, 3	" "	" 1¼		"	12.00
" 4, 4	" "	" 1½	and 2	"	20.00

Brass Cylinder.

No. 1, 2½	in. diam., suitable for	1		in. pipe,	$16.00
" 2, 2¾	" "	" 1	and 1¼	"	16.50
" 3, 3	" "	" 1¼		"	17.00
" 4, 4	" "	" 1½	and 2	"	33.00

The above Pumps are also made with Brackets to screw to wall or a board, at same price.

House Force Pumps.—With Cocks.— Single Acting.

Iron.

No. 1, 2	in. Cylinder, suitable for pipe 1	and 1¼ in. cal.	$18.00
" 2, 2½	" " "	1½ "	19.00
" 3, 3	" " "	1¼ and 1½ "	22.00

If without Cock and with Spout or Coupling, deduct $2.00.

Brass.

No. 1, 2	in. Cylinder, suitable for pipe 1	in. calibre	$35.00
" 2, 2½	" " "	1¼ "	37.00
" 3, 3	" " "	1½ "	42.00

Metalic Valves, extra, $4.00.
If without Cock and with Spout, deduct $5.00.

Double-Acting Suction and Force Pump.—On Plank.

With Crank and With Tight and Loose Pulleys for Power.

The Brass Pumps are always furnished with Iron Air Chambers, unless otherwise ordered. When Brass Air Chambers are ordered we charge additional cost of material only.

Length of Stroke seven inches.

SIZES AND PRICES.

	Iron	Brass.
No. 1, 2¼ in. calibre, for 1 or 1¼ in. pipe.	39.00	51.00
" 2, 2½ " " 1¼ "	41.00	56.00
" 3, 3 " " 1¼ " 1½ "	45.00	62.00
" 4, 3½ " " 1¼ " 2 "	51.00	81.00
" 5, 4 " " 1¼ " 2 "	68.00	114.00

Cock on side discharge, $2.50 extra.

Boiler Force Pumps.

Small Sizes.

Diameter of Piston, 1¼ inch	$10.00
" " 1½ "	15.00

Large Sizes.

Diameter of Piston, 2 inch	$ 22.00
" " 2½ "	30.00
" " 3 "	35.00
" " 4 "	60.00
" " 6 "	100.00

Plumbers' Force or Pressure Pumps.

For Clearing Obstructed Pipes.

This cut represents one of our Plumbers' Force or Pressure Pumps with Cylinder and working parts of cast brass, for clearing waste pipes that have become choked or obstructed.

Price, each $14.00

Gas Drip and Hydraulic Pressure Pumps.

Gas Drip Pump, for Gas Companies' Use.

Designed particularly for pumping gas drips. Fitted for attaching to ¾ or 1 inch wrought Iron Pipe.

Price $10.00

Garden Pump.

Fig. 375.—Double-Acting Hydropult.

Price, complete, with Suction Hose and Discharge pipe..... $9.00

Rotary Pumps.

Hand. **Power.**

SIZES AND PRICES.

SIZE	Iron.	Bronze.	With Bronze Cams Only.	FOR POWER. Iron.	FOR POWER. Brass.
No. 1, suitable for 1¼ in. pipes, price,	19.00	41.00	32.00	26.00	45.00
" 2, " 1½ " "	21.00	44.00	36.00	31.00	55.00
" 3, " 1½ " "	25.00	50.00	42.00	38.00	63.00
" 4, " 2 " "	34.00	64.00	50.00	48.00	75.00
" 5, " 2 " "	39.00	72.00	57.00	54.00	90.00

These Pumps when used for hot liquids, should, when permissable, be made with Iron Case and Bronze Cams, in order to provide against the unequal expansion of these parts.

No. 1, will discharge 13½ gallons per minute with 100 Revolutions
" 2, " " 14½ " " " 100 "
" 3, " " 17 " " " 100 "
" 4, " " 26 " " " 100 "
" 5, " " 35 " " " 100 "

Larger size Power Rotary Pumps furnished if required.

Pump Cylinders.

Screw Attachment.

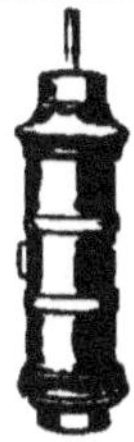

Diam.	10 in. Long	12 in. Long	14 in. Long.
2 in.	3.75	5.50	5.75
2¼ "	4.00	5.75	6.00
2½ "	4.35	6.00	6.50
2¾ "	4.70	6.50	7.00
3 "	5.00	7.00	7.50
3¼ "	5.30	7.50	8.00
3½ "	5.60	8.00	8.50
3¾ "	5.90	8.50	9.00
4 "	6.50	9.25	10.00

All sizes Brass Cylinders in stock.

Bolt Attachment.

With Flanges.

Diam.	16 in. Long.	18 in. Long.	20 in. Long.
2 in.	6.00	7.00	7.50
2¼ "	6.50	7.50	8.00
2½ "	7.00	8.00	8.50
2¾ "	7.50	8.50	9.00
3 "	8.00	9.00	9.50
3¼ "	8.50	9.25	10.25
3½ "	9.00	9.50	11.25
3¾ "	9.50	10.00	11.50
4 "	10.50	10.50	12.50

Without Flanges.

Red Jacket Pump.

The Working Part of Which can be Removed for Repairs Without Disturbing Pump.

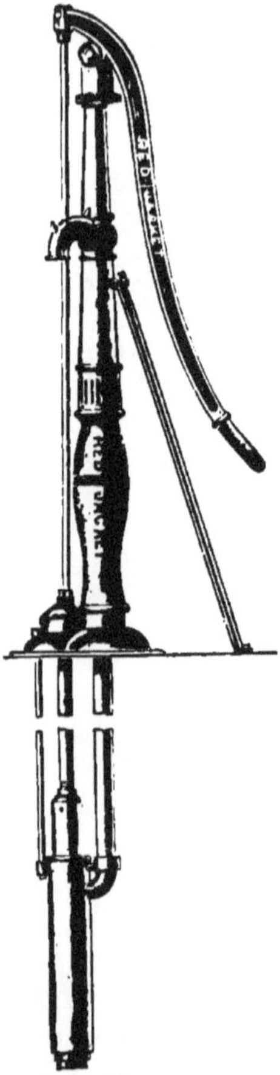

Fig. 12.

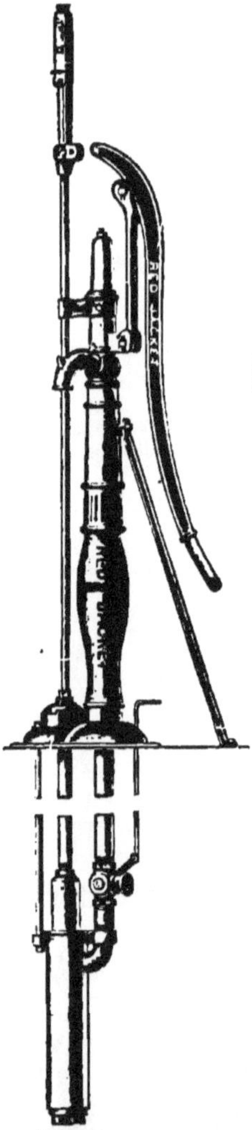

Fig. 13.

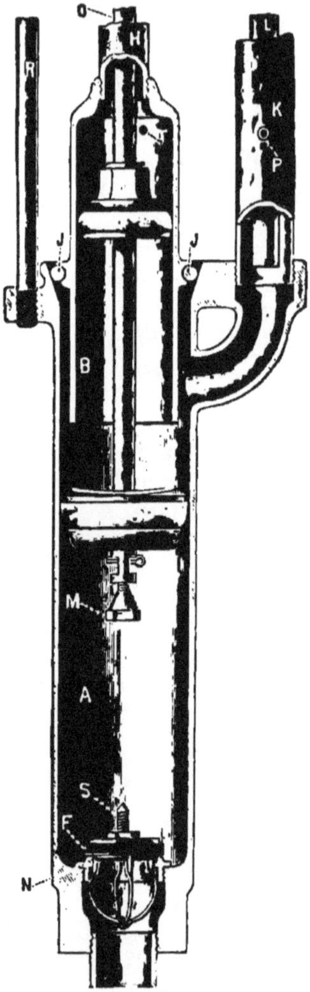

Fig. 14.

Sectional Views of Cylinders for Figures 12 and 13.

A—Lower Cylinder.	H—Washer Pipe.	N—Brass Valve Seat.
B—Upper Cylinder.	J—Rubber Cylinder Ring.	O—Piston Rod.
D—Lower Bucket.	K—Discharge Pipe.	P—Brass-Lined Frost Vent.
E—Upper Bucket.	L—Air Chamber.	R—Stay Rod.
F—Poppet-Valve Rubber.	M—Valve Guard.	S—Brass Valve Stem.

RED JACKET

Deep Well Force Pumps.

Fig. 15.

Sectional View of Deep-Well Cylinders and Pipe Connecting, Showing Wood Rod and Patent Malleable Couplings.

(See Fig. 14 for explanation of letters.)

How To Order.

1. Always give Number of Pump.

2. Only mention the 3-way Cock when wanted.

3. On deep well Pumps always mention the windmill attachment if wanted.

4. Do not order any Pump for a deeper well than specified.

5. If Pump is ordered fitted complete, give exact depth of well and amount of water, also specify kind of pipe wanted.

Points to Remember.

1. All Pumps are supplied with brace, strainer, and spout-coupling for hose.

2. "Set Length" means the complete length of Pump below the platform.

3. All suction pipe, also all large pipe for deep-well Pumps, is not included in price of Pumps.

4. The deep-well Pumps should be used with plugged and reamed galvanized Pipe between cylinders.

5. The deep-well Pumps are fitted for and can only be used with wood rod and couplings listed.

6. Don't try to sell the Red Jacket without Pumps to show and fill orders with.

Fig. 16.

RED JACKET PUMPS

Hand Use.—Fig. 12.

For Wells Under 26 Feet Deep.

Order by Number.	Set Length.	Size Cylinder.	Size Suction.	Price.	3-Way Cock, Extra.
No. 50.........	5 feet.	3 in.	1¼ in.	$16.00	$3.50
" 60.	5 "	3½ "	1½ "	18.00	3.50
" 70.........	5 "	4 "	2 "	22.00	5.00

For Wells 26 to 30 Feet Deep.

Order by Number.	Set Length.	Size Cylinder.	Size Suction.	Price.	3-Way Cock, Extra.
No. 51.........	9 feet.	3 in.	1¼ in.	$18.00	$3.50
" 61.........	9 "	3½ "	1½ "	20.00	3.50
" 71.........	9 "	4 "	2 "	24.00	5.00

For Wells 30 to 34 Feet Deep

Order by Number.	Set Length.	Size Cylinder.	Size Suction.	Price.	3-Way Cock, Extra.
No. 52.........	12 feet.	3 in.	1¼ in.	$20.00	$3.50
" 62.........	12 "	3½ "	1½ "	22.00	3.50
" 72.........	12 "	4 "	2 "	26.00	5.00

Windmill Head.—Fig. 13.

For Wells Under 26 Feet Deep.

Order by Number.	Set Length.	Size Cylinder.	Size Suction.	Price.	3 Way Cock, Extra.
No. 55.........	5 feet.	3 in.	1¼ in.	$17.00	$3.50
" 65.........	5 "	3½ "	1½ "	19.00	3.50
" 75........	5 "	4 "	2 "	24.00	5.00

For Wells 26 to 30 Feet Deep.

Order by Number.	Set Length.	Size Cylinder.	Size Suction.	Price.	3-Way Cock, Extra.
No. 56........	9 feet.	3 in.	1¼ in.	$19.00	$3.50
" 66.........	9 "	3½ "	1½ "	21.00	3.50
" 76.........	9 "	4 "	2 "	26.00	5.00

For Wells 30 to 34 Feet Deep.

Order by Number.	Set Length.	Size Cylinder.	Size Suction.	Price.	3-Way Cock, Extra.
No. 57.........	12 feet.	3 in.	1¼ in.	$21.00	$3.50
" 67.........	12 "	3½ "	1½ "	23.00	3.50
" 77.........	12 "	4 "	2 "	28.00	5.00

Red Jacket Pumps.

FIG. 15 WILL GO IN A FIVE INCH HOLE.

For Wells from 35 to 125 Feet Deep.

No. 87 B has a 2½ inch lower detached brass lined cylinder, tapped for 1¼ inch suction pipe, and 2½ inch pipe above.

Length below platform, including lower cylinder, 7½ feet.

Price	$23.50
Three Way Cock, extra	3.50
Windmill Attachment, extra	1.00

For Wells from 75 to 300 Feet Deep.

No. 82 A has a 2 inch lower detached brass lined cylinder, tapped for 1¼ inch suction pipe, and 2 inch pipe above.

Length below platform, including lower cylinder, 7½ feet.

Price	$21.50
Three-way Cock, extra	3.50
Windmill Attachment	1.00

Wood Rod and Couplings for Deep Well Pumps.

White Ash Rod, 12 feet lengths, per foot	$ .03
Couplings, Plain, per set	.20
Couplings, Tinned, per set	.30

Specify Plain or Tinned when ordering Couplings.

For fuller description of Red Jacket Pumps and Repair List, send for Circular.

Hydraulic Ram.

This cut represents our Improved Hydraulic Ram for supplying running water to Dwellings, Plantations, Factories, Railroad Stations, etc. The most complete article of the kind made.

No.	Quantity of Water furnished per Min. to Ram.	Length of Pipes.		Size of Pipes.		Price.
		Drive.	Dis'g.	Drive.	Dis'g.	
2	1 to 2 Gallons.	30 to 60 feet.	To where Desired	3/4 in. and	1/2 in.	$ 9.00
3	2 " 4 "	30 " 60		1 "	1/2 "	11.00
4	3 " 7 "	30 " 60		1 1/4 "	3/4 "	14.00
5	6 " 10 "	60		2 "	1 "	22.00
6	11 " 25 "	40 " 75		2 1/2 "	1 1/4 "	40.00
7	25 " 45 "	40 " 75		3 1/2 "	1 1/2 "	60.00
8	45 " 75 "	40 " 76		4 "	2 "	70.00

Hydrant and Street Washer.

Descriptive Circulars of these Goods Furnished on Application.

To Set in Ground, feet	3	4	5
3/4 in. Hydrants, each	10.60	11.00	11.50
1 " " "	14.30	14.75	15.30
3/4 " Street Washer, each	7.35	7.75	8.25
1 " " " "	11.10	11.70	12.00

Fire Hydrants.

We are Manufacturers' Agents for all the leading Hydrants used by the best Water Companies and Corporations throughout the United States. Circulars and particulars of the various kinds furnished on application.

STANDARD PRICE LIST FOR FIRE HYDRANTS.

Diameter of Pipe Connection.	Inside Diameter of Stand Pipe.	Diameter of Seat Ring or Gate Opening.	One 2 inch Nozzle.	One 2½ Nozzle.	Two 2½ Nozzles.	Three 2½ Nozzles.	Four 2½ Nozzles.	Six 2½ Nozzles.	One Steamer Nozzle.	One Steamer and One 2½ Nozzle.	One Steamer and Two 2½ Nozzles.	Frost Case, Standard Length.	For each 6 in. more or less than Standard Length of Stand Pipe, add to or deduct from List.	For each 6 in. more or less than Standard Length of Frost Case, add to or deduct from List.	Independent Nozzle Gate, each.
Inches.	Inches	Inches.													
2	3	2	$17.00	....	....	....	....	....	....	....	....	....	$ .45	....	....
3 or 4	4⅝	3	....	$28.00	....	....	....	....	....	....	....	$4.50	.60	$ .44	....
3 or 4	5¾	4	....	31.00	$33.00	$35.00	....	....	$33.00	$35.00	$37.00	5.00	.75	.50	$3.50
6	5¾	4	....	31.50	33.50	35.50	....	....	33.50	35.50	37.50	5.00	.75	.50	3.50
6	6¼	4½	....	....	....	37.50	$39.50	....	35.50	37.50	39.50	5.60	.80	.58	3.50
4 or 6	7	5	....	....	40.50	42.50	....	....	40.50	42.50	44.50	6.50	.85	.70	3.75
6	8	6	....	....	50.00	52.00	54.00	....	50.00	52.00	54.00	7.50	1.00	.90	3.75
8	8	6	....	....	51.25	53.25	55.25	....	51.25	53.25	55.25	7.50	1.00	.90	3.75
8 or 10	10	8	....	....	....	....	....	$130.00	....	....	....	11.00	2.25	1.30	4.50

The above prices are based on our standard length, viz.: five feet from ground surface to bottom of connecting pipe. Frost-cases are furnished if wanted.

Price List of Secondary Gates for Hydrants.

Subject to Same Discount as Hydrants.

3 inch Connection with 3 inch Valve	$ 9.25
4 " " " 4 " "	12.50
4½ " " " 5 " "	15.50
5 " " " 5 " "	15.50
6 " " " 6 " "	18.50
8 " " " 8 " "	27.00

Area of 2 inch Ring	3.141
" 2½ "	4.908
" 3 "	7.068
" 4 "	12.566
" 4½ "	15.904
" 5 "	19.635
" 5½ "	23.758
" 6 "	28.274
" 8 "	50.265
" 10 "	78.54

Patent Electro-Plated Well Point.

The accompanying cut is an illustration of an Improved Well Point, *A*, showing the openings (for admitting water) before the wire cloth or Perforated Brass is soldered on 1¼ inch by ¾ inch holes. *B*, showing section after wire cloth has been soldered on. *C*, showing section after both wire cloth and perforated brass have been soldered on. These 1¼ Points (calibre) are Electro-Plated with PURE BLOCK TIN inside and outside *before* receiving the coverings, there being no *raw iron* to *corrode or rust* or cause an *unpleasant taste* to the water.

1	inch	calibre,	price	each	....	$ 1.50
1¼	"	"	"	"	....	1.50
1½	"	"	"	"	....	2.50
2	"	"	"	"	....	6.00
2½	"	"	"	"	....	9.00
3	"	"	"	"	....	12.00
4	"	"	"	"	....	35.00

The 1¼ Point is 24½ inches long and has twenty-four 1¼x¾ openings. This is equal to 536 one-fourth inch drilled holes (commonly in Gas Pipe Point,) or equal to 342 five-sixteenth holes, or 239 three eighth drilled holes. So it is plain to be seen that the capacity of this Point is three to one greater than in any other in use, not barring length.

Fountains.

We also have designs of all the latest and finest patterns of Fountains, and can furnish them to order on short notice. Also Fountain Jets of every description.

CAST IRON DRAIN, WATER AND SOIL PIPE, AND FITTINGS.

All Pipe and Fittings Furnished Tarred, Unless Otherwise Ordered.

IN FIVE FOOT LENGTHS.

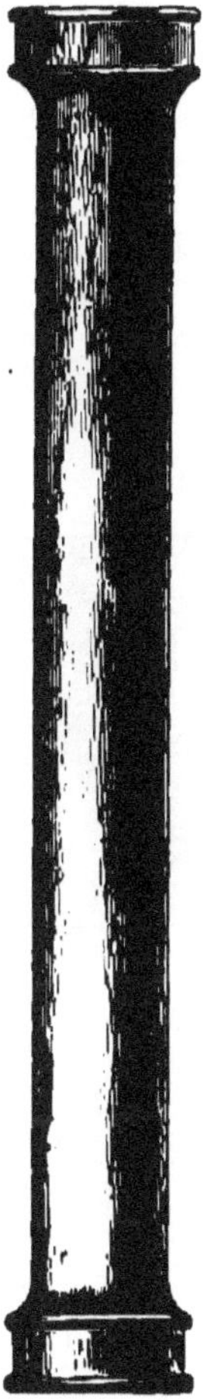

Double Hub.

Single Hub.

Single Hub.

	Standard.	Extra Heavy.
2 inches	$ .24	$ .35
3 "	.30	.55
4 "	.40	.75
5 "	.50	1.00
6 "	.60	1.20
7 "	1.00	1.75
8 "	1.25	2.25
10 "	2.00	3.00
12 "	3.00	4 00

Double Hub.

	Standard.	Extra Heavy.
2 inch, per Length	$1.50	$2.05
3 " "	1.80	3.05
4 " "	2.30	4.05
5 " "	2.80	5.30
6 " "	3.30	6.30
8 " "	7.25	12.75

Average Weight of Extra Heavy Cast Iron Pipe.

PER FOOT.

2 in.	3 in.	4 in.	5 in.	6 in.	7 in.	8 in.
5½ lbs.	9½ lbs.	13 lbs.	17 lbs.	20 lbs.	27 lbs.	33½ lbs.

Price for Enameling Pipes and Fittings, Net.

SIZE	2	3	4	5	6
Pipe, per Length,	1.40	1.50	1.60	1.70	1.80
Fittings, each	.50	.60	.75	.85	1.00

Quarter Bends.

	Standard.	Extra Heavy.
2 inch	$ .40	$ .50
3 "	.55	.70
4 "	.75	1.10
5 "	1.00	1.35
6 "	1.20	1.75
7 "	2.25	3.00
8 "	3.00	4.00
10 "	4.00	5.00
12 "	6.00	8.00

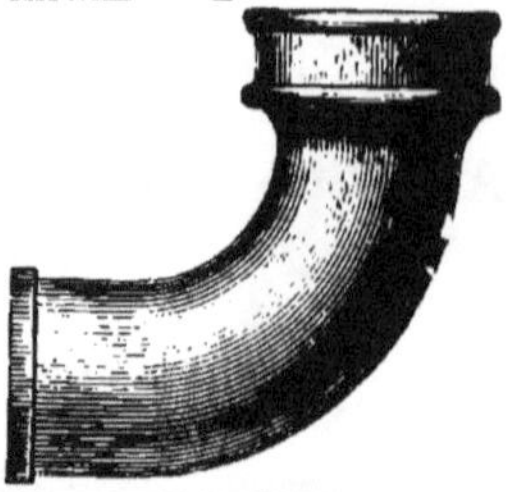

Long Bends.

18 Inches in Clear.

	Standard.	Extra Heavy.
4 inch ..	$1.50	$2.25
5 " ..	2.25	3.00
6 " ..	2.50	3.50

Quarter Bends—With Outlet on Side.

	Standard.	Extra Heavy.
2 in., with 2 in. Outlet, price, each	$ .90	$1.00
3 " 2 " "	1.05	1.20
4 " 2 " "	1.25	1.60
4 " 3 " "	1.35	1.70
5 " 2 " "	1.50	1.85
5 " 3 " "	1.60	1.95
6 " 2 " "	1.70	2.25
6 " 3 " "	1.80	2.35
6 " 4 " "	1.95	2.50

Quarter Bends—With Outlet in Heel.

	Standard	Extra Heavy.
2 in., with 2 in. Connection	$ .90	$1.00
3 " 2 "	1.05	1.20
4 " 2 "	1.25	1 60
4 " 3 "	1.35	1.70
5 " 2 "	1.50	1.85
5 " 3 "	1.60	1.95
6 " 2 "	1.70	2.25
6 " 3 "	1.80	2.35
6 " 4 "	1.95	2.50

Quarter Bends—Double Hub.

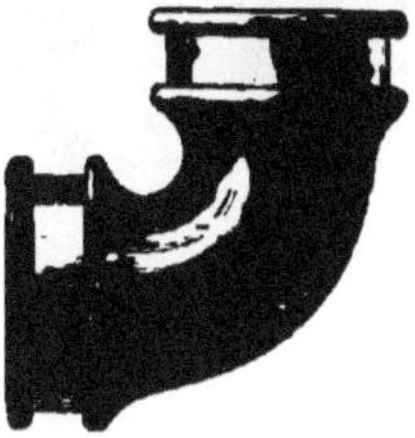

	Standard	Extra Heavy.
2 inch	$ .70	$ .80
3 "	.85	1.00
4 "	.95	1.40
5 "	1.30	1.65
6 "	1.50	2.05

Sixth Bends.

	Standard.	Extra Heavy.
2 inch	$ .40	$ .50
3 "	.55	.70
4 "	.75	1.10
5 "	1.00	1.35
6 "	1.20	1.75
7 "	2.25	3.00
8 "	3.00	4.00
10 "	4.00	5.00

Eighth Bends.

	Standard.	Extra Heavy.
2 inch	$ .40	$ .50
3 "	.55	.70
4 "	.75	1.10
5 "	1.00	1.35
6 "	1.20	1.75
7 "	2.25	3.00
8 "	3.00	4.00
10 "	4.00	5.00
12 "	6.00	8.00

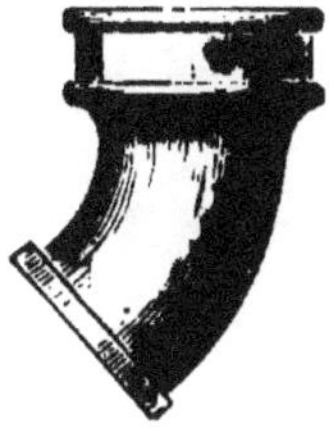

Sixteenth Bends.

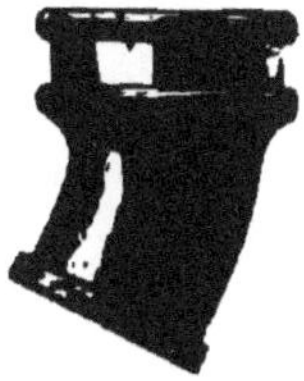

	Standard.	Extra Heavy.
2 inch	$ .40	$ .50
3 "	.55	.70
4 "	.75	1.10
5 "	1.00	1.35
6 "	1.20	1.75
8 "	3.00	4.00

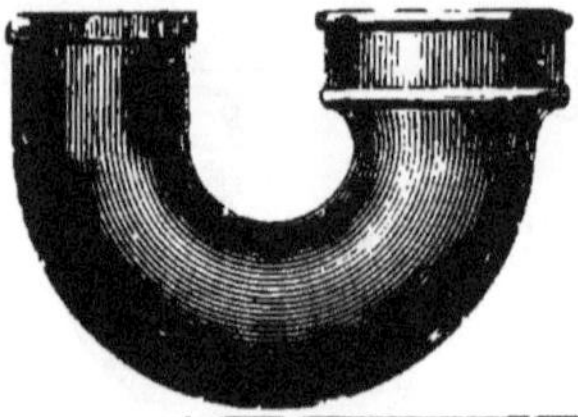

Return Bends.

	Standard.	Extra Heavy.
2 inch	.65	.75
3 "	.85	1.10
4 "	1.25	1.75
5 "	2.00	2.75
6 "	3.00	4.00

T Branches.

		Standard.	Extra Heavy.
2x2 inch	Each	.60	.80
3x3, 3x2 inch	"	.80	1.25
4x4, 4x3, 4x2 inch	"	1.20	1.60
5x5, 5x4, 5x3, 5x2 inch	"	1.60	2.25
6x6, 6x5, 6x4, 6x3, 6x2 inch	"	2.00	3.25
7x7, 7x6, 7x5, 7x4, 7x3, 7x2 inch	"	4.00	6.00
8x8, 8x6, 8x5, 8x4, 8x3, 8x2 inch	"	5.00	8.00
10x10, 10x8, 10x6, 10x4 inch	"	7.00	11.00

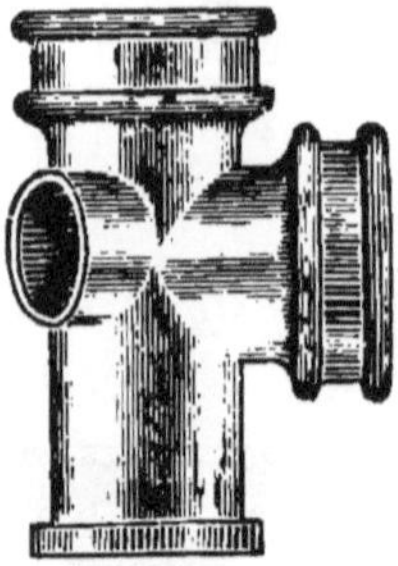

T Branch with Inlet.

Right or Left.

Add to regular list for 2 inch Inlet	$ .50
" " " " 3 " "	.60
" " " " 4 " "	.75

Long T Branches.

Twenty-four Inches in Clear.

	Standard.	Extra Heavy.
4x4 inch	2.50	3.50

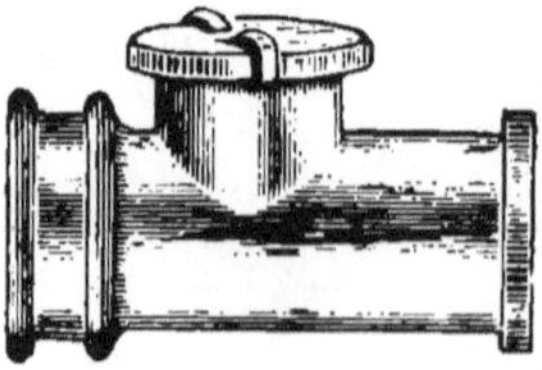

Hand Hole Ts.

	Standard.	Extra Heavy.
4x4 inch	1.25	1.75
5x4 "	1.75	2.25
6x5 "	2.25	3.00

Vent Branches for Back Air Pipe.

		Standard.	Extra Heavy.
2x2	Each,	.80	1.25
3x2	"	1.25	1.75
4x2	"	1.50	2.00
5x2, 5x3, 5x4	"	2.00	2.75
6x2, 6x3, 6x4	"	3.00	4.00

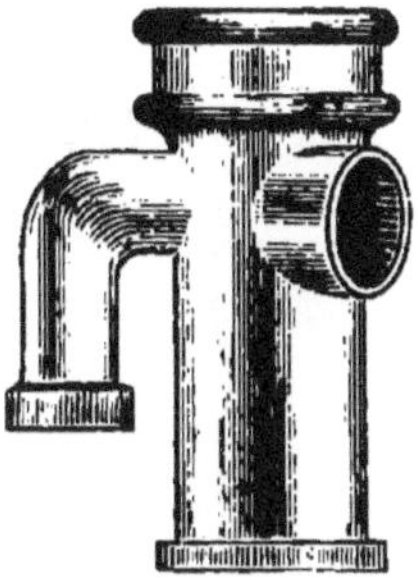

Vent Branches for Back Air Pipe.

With Right or Left Inlet.

Add to regular list for 2 inch Inlet	$ .50
" " " 3 " "	.60
" " " 4 " "	.75

Crosshead Branches.

		Standard	Extra Heavy.
2x2	Each	1.00	1.25
3x3, 3x2	"	1.25	1.60
4x4, 4x3, 4x2	"	1.65	2.00
5x5, 5x4, 5x3, 5x2	"	2.25	3.00
6x6, 6x5, 6x4, 6x3. 6x2	"	3.00	4.00
7x7, 7x6, 7x5, 7x4	"	5.50	7.00
8x8, 8x6, 8x5, 8x4	"	6.00	9.00
10x10. 10x8. 10x6	"	9.00	14.00

Y Branches.

		Standard	Extra Heavy.
2x2	Each	.60	.80
3x3, 3x2	"	.80	1.25
4x4, 4x3, 4x2	"	1.20	1.60
5x5, 5x4, 5x3, 5x2	"	1.60	2.25
6x6, 6x5, 6x4, 6x3, 6x2	"	2.00	3.25
7x7, 7x6, 7x5, 7x4, 7x3	"	4.00	6.00
8x8, 8x6, 8x5, 8x4	"	5.00	8.00
10x10	"	7.00	11.00

Y Branch, with Inlet.

Right or Left.

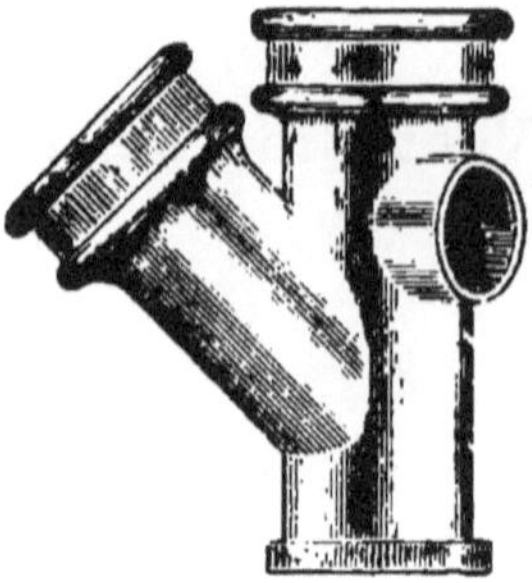

Add to regular list for 2 inch Inlet			$ .50
" " " 3 " "			.60
" " " 4 " "			.75

Long Y Branches.

	Standard.	Extra Heavy.
4x4, 24 inch in the clear	2.50	3.50

Double Y Branches.

		Standard.	Extra Heavy.
2x2	Each	1.00	1.25
3x3, 3x2	"	1.25	1.60
4x4, 4x3, 4x2	"	1.65	2.00
5x5, 5x4, 5x3, 5x2	"	2.25	3.00
6x6, 6x5, 6x4, 6x3, 6x2	"	3.00	4.00
7x7, 7x6, 7x5, 7x4	"	5.50	7.00
8x8, 8x6, 8x5, 8x4	"	6.00	9.00
10x10, 10x8, 10x6, 10x4	"	9.00	14.00

Heavy Y Branches.

		Standard.	Extra Heavy
2x2	Each	.60	.80
3x3, 3x2	"	.80	1.25
4x4, 4x3, 4x2	"	1.20	1.60
5x5, 5x4, 5x3, 5x2	"	1.60	2.25
6x6, 6x5, 6x4, 6x3, 6x2	"	2.00	3.25
7x7, 7x6, 7x5, 7x4, 7x3, 7x2	"	4.00	6.00
8x8, 8x7, 8x6, 8x5, 8x4, 8x3, 8x2	"	5.00	8.00

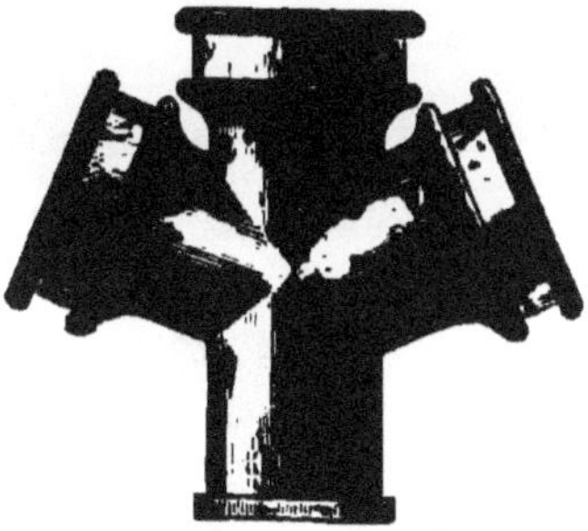

Double Half Y Branches.

		Standard.	Extra Heavy.
2x2	Each	1.00	1.25
3x2, 3x3	"	1.25	1.60
4x4, 4x3, 4x2,	"	1.65	2.00
5x5, 5x4, 5x3, 5x2	"	2.25	3.00
6x6, 6x5, 6x4, 6x3, 6x2	"	3.00	4.00
7x7, 7x6, 7x5, 7x4, 7x3, 7x2	"	5.50	7.00
8x8, 8x7, 8x6, 8x5, 8x4, 8x3, 8x2	"	6.00	9.00

T Y Branches.

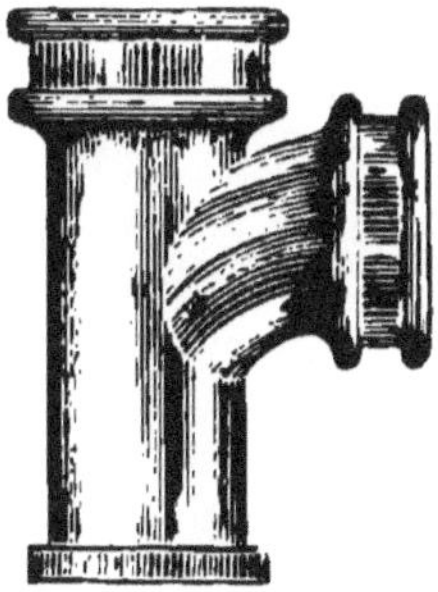

		Standard.	Extra Heavy.
2x2,	Each	.60	.80
3x3, 3x2	"	.80	1.25
4x4, 4x3, 4x2	"	1.20	1.60
5x5, 5x4, 5x3, 5x2	"	1.60	2.25
6x6, 6x5, 6x4, 6x3, 6x2, in.	"	2.00	3.25

T Y Branch with Inlet.

Right or Left.

Add to regular list for	2 in. inlet		$	.50
" "	3	"		.60
" "	4	"		.75

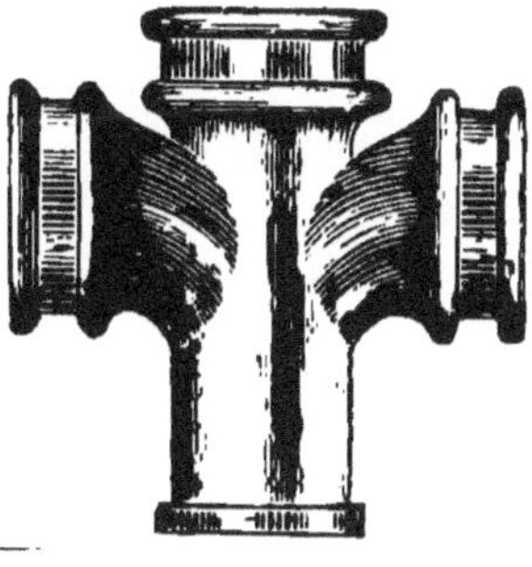

Double T Y Branches.

		Standard.	Extra Heavy.
2x2	Each.	1.00	1.25
3x3, 3x2	"	1.25	1.60
4x4, 4x3, 4x2	"	1.65	2.00
5x5, 5x4, 5x3, 5x2	"	2.25	3.00
6x6, 6x5, 6x4, 6x3, 6x2 in	"	3.00	4.00

Offsets.

					Standard.	Extra Heavy.
2	inch,	to offset	2	inch.	.40	.60
2	"	"	4	"	.50	.90
2	"	"	6	"	.60	1.00
2	"	"	8	"	.70	1.10
2	"	"	10	"	.80	1.20
2	"	"	12	"	.85	1.25
2	"	"	14	"	1.00	1.45
2	"	"	16	"	1.15	1.60
2	"	"	18	"	1.25	1.75
2	"	"	20	"	1.40	1.95
3	"	"	4	"	.75	1.10
3	"	"	6	"	.80	1.20
3	"	"	8	"	.90	1.35
3	"	"	10	"	.95	1.40
3	"	"	12	"	1.00	1.45
3	"	"	14	"	1.25	1.70
3	"	"	16	"	1.40	1.90
3	"	"	18	"	1.50	2.00
3	"	"	20	"	1.65	2.25
4	"	"	4	"	.85	1.25
4	"	"	6	"	1.00	1.40
4	"	"	8	"	1.15	1.50
4	"	"	10	"	1.25	1.60
4	"	"	12	"	1.40	1.80
4	"	"	14	"	1.65	2.00
4	"	"	16	"	1.80	2.25
4	"	"	18	"	2.15	2.80
4	"	"	20	"	2.25	3.00
5	"	"	4	"	1.40	1.80
5	"	"	6	"	1.60	2.00
5	"	"	8	"	1.80	2.25
5	"	"	10	"	1.90	2.40
5	"	"	12	"	2.00	2.50
5	"	"	16	"	2.40	3.00
6	"	"	4	"	2.00	3.00
6	"	"	6	"	2.25	3.25
6	"	"	8	"	2.40	3.50
6	"	"	10	"	2.60	3.75
6	"	"	12	"	2.75	4.00

Offsets with Two-inch Outlets.

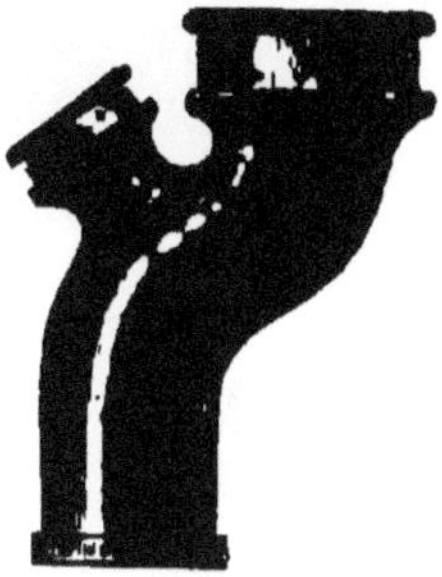

	Standard.	Extra Heavy.
4 inch, to offset 4 inch	1.35	1.75
4 " " 6 "	1.50	1.90
4 " " 8 "	1.65	2.00
4 " " 10 "	1.75	2.10
4 " " 12 "	1.90	2.30
4 " " 14 "	2.15	2.50
4 " " 16 "	2.30	2.75
4 " " 18 "	2.65	3.30
4 " " 20 "	2.75	3.50

Double Hubs.

	Standard.	Extra Heavy.		Standard.	Extra Heavy.
2 inches	.30	.40	6 inches	.80	1.15
3 "	.45	.55	7 "	1.40	2.50
4 "	.65	.75	8 "	2.50	3.50
5 "	.75	.90	10 "	3.50	5.50

Straight Sleeves.

	Standard.	Extra Heavy.		Standard.	Extra Heavy.
2 inches	.30	.40	6 inches	.80	1.15
3 "	.45	.55	7 "	1.40	2.50
4 "	.65	.75	8 "	1.50	3.50
5 "	.75	.90	10 "	2.50	4.50

Single Hubs.

	Standard.	Extra Heavy.		Standard	Extra Heavy.
2 inches	.25	.35	6 inches	.75	1.00
3 "	.35	.40	7 "	1.25	2.00
4 "	.40	.50	8 "	2.50	3.00
5 "	.60	.75	10 "	3.50	5.00

Reducers.

	Standard.	Extra Heavy.
To reduce from 3 to 2 inch	.45	.55
" " 4 " 2 "		
" " 4 " 3 "	.50	.60
" " 5 " 2 "		
" " 5 " 3 "	.70	.80
" " 5 " 4 "		
" " 6 " 2 "		
" " 6 " 3 "	.80	.90
" " 6 " 4 "		
" " 6 " 5 "		
" " 7 " 2 "		
" " 7 " 3 "	1.30	2.00
" " 7 " 4 "		
" " 7 " 5 "		
" " 8 " 2 "		
" " 8 " 3 "		
" " 8 " 4 "	1.60	2.20
" " 8 " 5 "		
" " 8 " 6 "		

Increasers.

	Standard.	Extra Heavy.
2x3 inch	.70	1.00
2x4 "	.75	1.10
2x5 "	.80	1.15
2x6 "	.85	1.30
3x4 "	.90	1.25
3x5 "	1.00	1.40
3x6 "	1.20	1.70
4x5 "	1.15	1.60
4x6 "	1.25	1.75
5x6 "	1.35	1.95
5x7 "	1.60	2.25
6x7 "	1.75	2.50
6x8 "	2.00	2.75

Pipe Plugs.

	Standard.	Extra Heavy.
2 inch	.15	.25
3 "	.25	.35
4 "	.30	.40
5 "	.35	.50
6 "	.50	.65
7 "	.90	1.25
8 "	1.20	1.50

Thimbles.

SIZE		3	4	5	6 in.
Plain	.15	.25	.30	.35	.45
Galvanized	.25	.40	.50	.60	.75

Thimbles.—With Cover.

Standard or Extra Heavy.

SIZE	2	3	4	5	6	8 in.
Price	.40	.50	.60	.70	.90	2.25

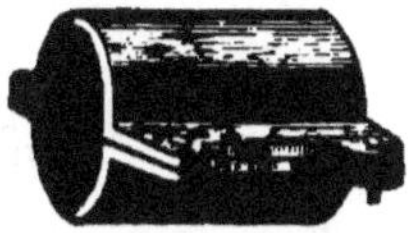

Pipe Bands.

SIZE	2	3	4	5	6 in.
Plain	.45	.55	.70	1.00	1.40
With outlet	.75	.90	1.10	1.45	1.90

Ventilating Caps.

SIZE	2	3	4	5	6 in.
Price, with Spigot	.40	.60	.80	1.10	1.50
" " Hubs	.70	.90	1.10	1.40	1.80

Saddle Hubs.

	T. Standard.	T. Extra Heavy	Half Y and Y. Standard.	Half Y and Y. Extra Heavy
2x2	.30	.40	.35	.45
3x3, 3x2	.50	.65	.55	.70
4x4, 4x3, 4x2	.60	.80	.70	.90
5x5, 5x4, 5x3, 5x2	.75	1.00	.90	1.15
6x6, 6x5, 6x4, 6x3, 6x2	1.10	1.40	1.25	1.55
7x6, 7x4, 7x2	1.40	2.00	......	....
8x6, 8x4, 8x2	1.50	2.25	2.00	3.00

S Traps.

	Standard.	Extra Heavy.
2 inch	.80	1.25
3 "	1.25	2.00
4 "	1.75	2.75
5 "	3.00	4.00
6 "	4.00	5.50

Three-Quarters S Traps.

	Standard.	Extra Heavy.
2 inch	.80	1.25
3 "	1.25	2.00
4 "	1.75	2.75
5 "	3.00	4.00
6 "	4.00	5.50

Half S Traps.

	Standard.	Extra Heavy.
2 inch	.80	1.25
3 "	1.25	2.00
4 "	1.75	2.75
5 "	3.00	4.00
6 "	4.00	5.50

Traps Without Hand Openings.

	Standard.	Extra Heavy.
4 in. Half S	1.75	2.75
4 " Three-quarter S	1.75	2.75
4 " Full S	1.75	2.75

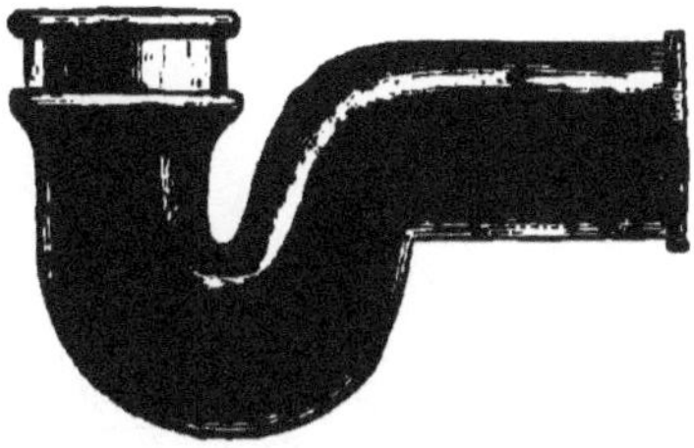

Running Traps.

	Standard.	Extra Heavy.
2 inch	.80	1.25
3 "	1.25	2.00
4 "	1.75	2.75
5 "	3.00	4.00
6 "	4.00	5.50
7 "	7.00	9.00
8 "	9.00	12.00

Traps.

With 2 inch Outlet in Heel.

	Standard.	Extra Heavy.
4 inch S, Half S, or ¾ S	2.25	3.25

Traps.

With 2 inch Outlet in Right or Left Side.

	Standard.	Extra Heavy.
4 inch **S** or Half **S**	2.25	3.25

Traps.

With 2 inch Outlet for Vent Pipe.

	Standard.	Extra Heavy.
2 inch............................	1.30	1.75
3 "	1 75	2.50
4 "	2.25	3.25
5 "	3.50	4.50
6 "	4.50	6.00

Running Traps.

With Hub for Vent Pipe.

	Standard	Extra Heavy.
4 inch....................	1.75	2.75
5 "	3.00	4.00
6 "	4.00	5.50
8 "	9.00	12.00

Running Traps.

With Hubs for Double Vent.

	Standard.	Extra Heavy.
4 inch..........	2.75	3.75
5 "	4.00	5.00
6 "	5.00	6.50
8 "	11.00	14.00

Trap Covers.

2 inch	$ .12
3 "	.16
4 "	.20
5 "	.30
6 "	.40
8 "	.60

Roof Irons.

2 inch	$ .90
3 "	1.15
4 "	1.30
5 "	1.50
6 "	1.80

"Dandy" Cleanouts.

Standard or Extra Heavy.

2 inch	$1.50
3 "	2.00
4 "	2.50
5 "	3.00
6 "	3 50

Soil Pipe Testing Plugs.

2 inch, each	$1.00
3 " "	1.25
4 " "	1.50
5 " "	2.00
6 " "	2.50

Leader Pipes.

WITH OR WITHOUT LUGS.

Length, 4 Feet 6 Inches.

2 inch	each	$3.00
4 "	"	4.00
5 "	"	5.00
6 "	"	6.00
4 " 8 feet long	"	8.00

Corner Mangers.

Plain,	each	$2.50
Galvanized	"	5.00

Extension Service Boxes.

Diameter of Box, 2½ Inches.

92 D	2 ft.	to 3 ft. 6 in.
93 D	3 "	" 4 "
93 E	3 "	" 4 " 6 "
94 D	3 " 6 in.	" 4 " 9 "
94 E	3 " 6 "	" 5 "
95 E	4 "	" 5 " 6
100 E	4 " 6	" 6 "

Extension Sections.

151	Incr'ng Length of Service Box	9½ in.
152	" " "	16½ "
153	" " "	28 "

Valve Boxes.

5¼ Inch Shaft.

AA	extends	1 ft. 10	to 2 ft. 4 in.
A	"	2 " 4	" 3 " 4 "
B	"	3 "	4 "
C	"	3 " 6	to 4 " 6 in.
CC	"	4 "	5 "
D	"	3 " 6	to 5 " 6 "
DD	"	4 "	" 6 "
E	"	5 "	" 6 "
F	"	5 "	" 7 "

Above are the Standard Sizes of these goods, which we keep constantly on hand, prices of which will be given on application.

Special sizes furnished to order.

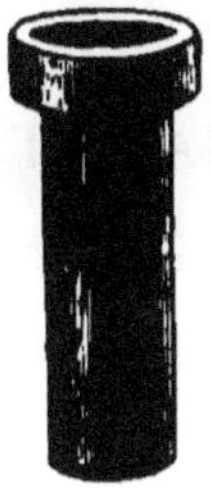

Vitrified Salt Glazed Sewer Pipe.

Adopted by Eastern and Western Manufacturers, Jan. 20, 1887.

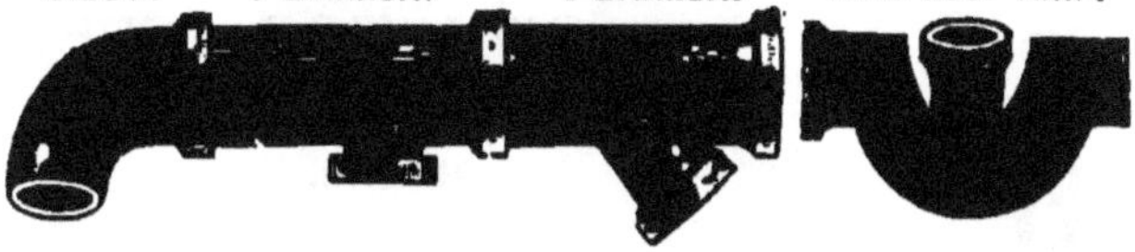

Caliber of Pipe.	Price per foot.	Elbows and Curves 2 ft. Long or less, each.	Branches, 1 ft. long, each	Branches, 2 ft. long, each.	Branches, 3 ft. ong, each.	Each Inlet in addition to Single Branches.	Slants per 1 foot or less. Long side.	Traps, each.	Weight of pipe per Foot.	Area in inches.	Caliber of pipe.
2 in.	.14	.40	.49	.63	.77	.35	....	1.00	6 lbs.	3.141	2 in
3 "	.16	.50	.56	.72	.88	.40	....	1.50	8 "	7.068	3 "
4 "	.20	.65	.70	.90	1.10	.50	.30	2.00	10 "	12.566	4 "
5 "	.25	.85	.88	1.13	1.38	.63	.37½	2.50	12 "	19.635	5 "
6 "	.30	1.10	1.05	1.35	1.65	.75	.45	3.50	16 "	28.274	6 "
8 "	.45	1.80	1.58	2.03	2.48	1.13	.67½	5.50	22 "	50.265	8 "
9 "	.55	2.25	1.93	2.48	3.03	1.38	.82½	6.50	26 "	63.617	9 "
10 "	.65	2.75	2.28	2.93	3.58	1.63	.97½	7.50	32 "	76.539	10 "
12 "	.85	3.50	2.98	3.83	4.68	2.13	1.27½	10.00	45 "	113.09	12 "
15 "	1.25	4.75	4.38	5.63	6.88	3.13	1.87½	15.00	63 "	176.71	15 "
18 "	1.70	6.50	5.95	7.65	9.35	4.25	2.55	....	84 "	254.46	18 "
20 "	2.25	7.50	7.88	10.13	12.38	5.63	3.37½	....	99 "	314.16	20 "
24 "	3.25	11.00	11.38	14.63	17.88	8.13	4.87½	....	120 "	452.39	24 "

Traps, without Hand Hole, shall be same price; but with more than one Hand Hole, the additional shall be charged the same as Inlets in Branches.

Two Piece Traps, 9 inches in diameter, or larger, to be charged eight times the price of one foot of pipe for each pair.

Pipe of all Sizes may be furnished with sockets or rings, and ring pipe sold without rings may be sold at a discount of ten per cent. from bill.

Increasers shall have socket on small end. **Reducers,** socket on large end, and shall be charged at double the price of two feet of pipe, size of larger end.

Each piece of **Channel** or **Split Pipe** shall be charged as three-fifths of a whole pipe.

Stoppers or **Plugs** for closing pipe, one-third of one foot of pipe of the size in which it is used.

Slop Bowls with strainer burned in, 3"x12", $2.50. **Closet Bowls,** 6"x12", $2.50.

Cellar Traps, size, 9", with 4" outlet, $5.00; size 12", with 4" outlet, $7.00.

Grease Traps, size 12", with 4" inlet and outlet, $7.00: size 15", with 6" inlet and outlet, $10.00; size 18", with 6" inlet and outlet $12.00.

We also can furnish at low prices Fire Brick of all kinds, Fire Clay Chimney Tops and Wind Guards, Flue or Chimney Linings and Agricultural Drain Tile.

Circulars and Price List on application.

AUSTIN'S PATENT CORRUGATED CONDUCTOR PIPE.

Round.

SIZE	2	3	4	5	6 in.
Galvanized, per foot	.12	.15	.20	.25	.30
Tin, "	.09	.12	.15	.20	.25

Square.

	Galvanized.	Tin.
Sizes, 1¾x2¼, equal to 2 inch Round, per foot	.12	.09
" 2⅛x2¾, " 3 " "	.15	.12
" 2⅝x3⅝, " 4 " "	.20	.15
" 3¼x4½, " 5 " "	.25	.20
" 3⅝x5¼, " 6 " "	.30	.25

Elbows and Shoes.

SIZE	2	3	4	5	6 in.
Elbows, Galvanized, round, each	.25	.30	.40	.50	.60
Shoes, " " "	.30	.36	.48	.60	.72
Elbows, " square, "	.30	.36	.48	.60	.72
Shoes, " " "	.40	.48	.60	.72	.84
No. 1, Tin, per dozen	2.25	2.50	2.75	3.50	4.25
" 2, " "	2.75	3.00	3.25	4.00	4.75
" 3, " "	3.25	3.50	3.75	4.50	5.25

In ordering Elbows and Shoes, purchasers will please state the angles required, referring to the numbers in the above cuts.

Special quotations furnished for conductor made of copper or other sheet metals.

Spiral Seam Galvanized House Leader, Ventilating, Air and Blower Pipe, &c.

Manufactured in Lengths of 10 Feet and Less.

INSIDE DIAMETER...	2	2½	3	3½	4	5	6 in.
Galvanized, per ft.	.14	.17	.19	.21	.25	.30	.38
Tin, "	.09	....	.11	....	.15	.21	.28

Patent Adjustable Elbows.

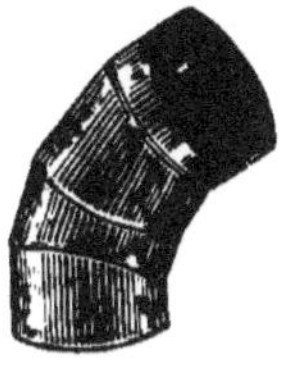

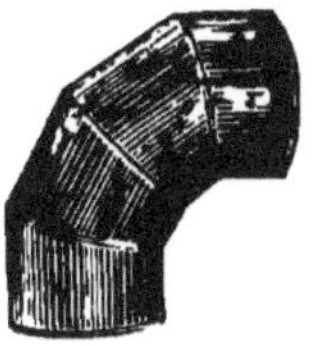

SIZE....	2	2½	3	4	5	6	7	8	9	10 in.
Galvanized, per dozen.......	1.60	2.00	2.40	3.00	3.60	5.40	7.20	9.60	10.00	12.50
Tin, per dozen...	1 30	1.80	1.80	2 25	3.50	5.00	6.50	8 00	10.00	11.50

Globe Ventilators.

SIZE.....	2	2½	2¾	3	3½	4	4½	5	5½	6	7	8	10	12 in.
Galv.Iron,Ea	1.00	1.00	1.00	1.50	1.50	1.75	2.00	2.50	2.85	3.40	4.00	4.65	5.75	6.75

SIZE.........	14	16	18	20	24	30	36	40	48	60 in.
Galv. Iron, Each.	13.00	20.00	27.00	33.00	40.00	65.00	120.00	180.00	240.00	360.00

If square or oblong base is required, it is charged extra.

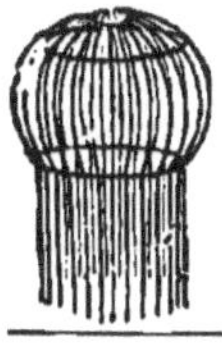

Galvanized Wire Conductor Strainers.

SIZE..............	2	3	4	5	6 in.
Per dozen........	2.50	3.00	3.50	4.00	4.50

Gas Pipe Hooks.

SIZE	¼	⅜	½	¾	1	1¼	1½	2 in.
Per 100	.40	.55	.60	.70	.95	1.20	1.30	1.60

Soil Pipe Hooks.

SIZE	2	3	4	5	6 in.
Per 100	1.50	2.00	2.75	3.50	4.50

Plumbers' Hooks.

SIZE	½	⅝	¾	1 in.
Each	.02	.02	.03	.04

Leader Hooks.

SIZE	2	3	4	5	6 in.
Tinned, each	.04	.05	.06	.09	.12
Black, "	.03	.04	.05	.06	.09

Austin's Leader Fastners, for Wood or Brick, per 100 sets..$7.00

Straps.

SIZE	¼	⅜	½	⅝	¾	1	1¼	1½ in.
Tinned, each	.01	.02	.02	.03	.03	.05	.07	.10

Lead Tacks, per lb .... $ .20

RUBBER HOSE.

We are the exclusive agents in Central New York for the **Gutta Percha and Rubber Manufacturing Co.**, one of the largest manufacturers of reliable Rubber Goods in the world. These goods will be found to be excellent in quality and reasonable in price. A complete Special Circular relating exclusively to them will be forwarded on application.

The 2-Ply Hose or Conducting Hose, is not calculated to stand much pressure.

The 3-Ply Hose is made to withstand a pressure of 75 lbs. to the square inch.

The 4 Ply Hose is made to withstand a pressure of 150 lbs. to the square inch.

Rubber Conducting Hose, 2-Ply.

INSIDE DIAM	½	¾	1	1¼	1½	1¾	2	2¼	2½	2¾	3 in.
Per foot	.20	.25	.33	.42	.50	.58	.66	.75	.83	.92	.99
INSIDE DIAM	4	5	6	7	8	9	10 in				
Per foot	1.32	1.65	1.98	2.31	2.64	2.97	3.33				

Rubber Hydrant Hose, 3-Ply.

INSIDE DIAM	½	¾	1	1¼	1½	1¾	2	2¼	2½	2¾	3 in.
Per foot	.25	.30	.40	.50	.60	.70	.80	.90	1.00	1.10	1.20

Rubber Engine Hose, 4-Ply.

INSIDE DIAM	½	¾	1	1¼	1½	1¾	2	2¼	2½	2¾	3 in.
Per foot	.30	.37	.50	.62	.75	.87	1.00	1.12	1.25	1.37	1.50

All the above kinds of Hose are kept on hand in lengths of 25 and 50 feet, and these we do not cut.

Five and Six-ply Hose made at an advance of 25 and 50 per cent. respectively on Four-ply prices.

Rubber Steam Hose.

INS DIAM...........	½	¾	1	1¼	1½	1¾	2	2½	3 in.
Three-Ply, per ft..	.45	.54	.71	.85	1.02	1.18	1.34	1.66	2.00
INS. DIAM...........	½	¾	1	1¼	1½	1¾	2	2½	3 in.
Four-Ply, per ft...	.51	.67	.83	1.04	1.25	1.45	1.66	2.08	2.80
INS. DIAM...........	½	¾	1	1¼	1½	1¾	2	2½	3 in.
Five-Ply, per ft.....	.63	.83	1.03	1.30	1.56	1.81	2.07	2.60	3.50
INS. DIAM...........	½	¾	1	1¼	1½	1¾	2	2½	3 in.
Six-Ply, per ft.....	.76	1.00	1.24	1.56	1.87	2.17	2.49	3.12	4.20

Steam Hose served with Marlin at 10 per cent. advance on Price List.
For each additional Ply add 25 per cent. of Four-Ply Prices.
Larger sizes made when required.

Rubber Brewers' Hose.

INS. DIAM...........	½	¾	1	1¼	1½	1¾	2	2½	3 in.
Three-Ply, per ft..	.43	.51	.67	.85	1.02	1.18	1.34	1.66	2.00
INS. DIAM...........	½	¾	1	1¼	1½	1¾	2	2½	3 in.
Four-Ply, per ft....	51	.67	.83	1.04	1.25	1.45	1.66	2.08	2 80

Five and Six-Ply Brewers' Hose made at an advance of 25 and 50 per cent. respectively on Four-Ply prices.

Rubber Suction Hose.

On Spiral Brass Wire.

INS. DIAM....................................	¾	1	1¼	1½	1¾	2 in.
Per foot......................................	.77	1.00	1.25	1.65	2.10	2.50

On Spiral Tinned or Iron Wire.

INS. DIAM....................................	¾	1	1¼	1½	1¾	2 in.
Per foot......................................	.70	.90	1.15	1.50	1.90	2.30

Rubber Oil Hose.

INS. DIAM.	½	¾	1	1¼	1½	1¾	2	2½	3 in.
Three-Ply, per ft.	.43	51	.67	.85	1.02	1.18	1.34	1.66	2.00
INS. DIAM.	½	¾	1	1¼	1½	1¾	2	2½	3 in.
Four-Ply, per ft.	.51	.67	.83	1.04	1.25	1.45	1.66	2.08	2.80

Served with Marlin at 10 per cent. advance on list prices.
For each additional Ply add 25 per cent. of Four-Ply prices.
Larger sizes made when required.

Rubber Tubing.

Made with Walls $\frac{1}{8}$, $\frac{3}{32}$, and $\frac{1}{16}$ inch thick, and put up in boxes of convenient size, containing 50 to 100 feet each.

INS. DIAM.	$\frac{1}{8}$	$\frac{3}{16}$	$\frac{1}{4}$	$\frac{5}{16}$	$\frac{3}{8}$	$\frac{1}{2}$	$\frac{5}{8}$	$\frac{3}{4}$	1 in.
Per foot	.08	.12	.16	.18	.20	.25	.30	.35	.45

Rubber Belting.

WIDTH	2	2½	3	3½	4	4½	5	6	7	8	9 in.
3-Ply, per foot.	.17	.22	.26	.30	.34	.39	.43	.52	.60	.70	.80
4-Ply, "	.21	.26	.31	.37	.42	.47	52	.62	.73	.84	.95

WIDTH	10	11	12	13	14	15	16	18	20	22	24 in.
3-Ply, per foot.	.90	1.00	1.08	1.18	1.28	1.38	1.50	1.70	1.90	2.12	2.36
4-Ply, "	1.07	1.18	1.30	1.42	1.54	1.66	1.78	2.02	2.26	2.52	2.80

Heavy 5 and 6-Ply Belts made to order for purposes where great strength is required, (as a substitute for double leather,) at an advance of twenty-five and fifty per cent. on 4-Ply prices.

2-Ply Rubber Machine Belting.

For Agricultural Machines, Railway Belts and other Light Work.

WIDTH	1	1¼	1½	2	2½	3	3½	4 in.
Per foot	.07	.09	.11	.15	.18	.22	.26	.30

Endless Belts, of any width or length, made to order at ten days' notice at current list prices, with an additional charge for the joining, equal to the price of three feet of the Belt.

Rubber Lined Cotton Hose.—Seamless Woven.

Strong! Light! Cheap!

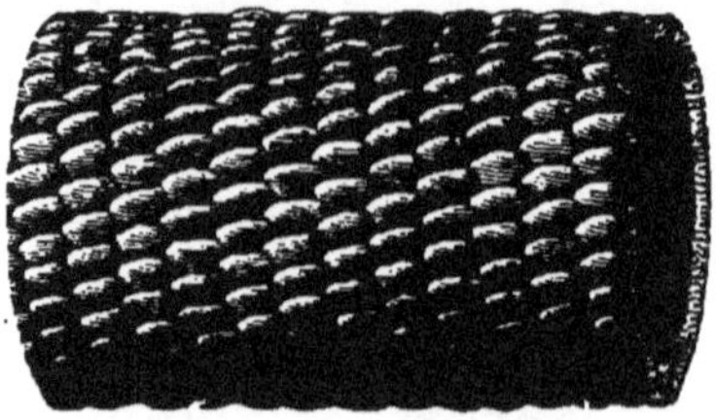

If given a reasonable chance to dry after being used, it will last many years. The ½ and ¾ inch has couplings attached.

SIZE INSIDE DIAM....................	½	¾	1	1¼	1½	2	2½ in.
Price, per foot, (50 feet lengths,)	.20	.25	.35	.45	.50	.60	.70

Linen Hose.—Unlined and Seamless.

SIZE INSIDE DIAM...........	1	1¼	1½	1¾	2	2¼	2½	3 in.
Standard, per foot..........	.20	.22	.25	.28	.30	.33	.35	.50
Trade, "	.15	.17	.20	......	.24	.26	.28	.40

Wire Bound Rubber Hose.

Warranted not to Kink.

List prices same as for Hose without wire.
Discounts furnished on application.

We can furnish Hose of every description for Fire Department and Factory use at the lowest prices.

We keep in stock a large assortment of Hose Reels. Circulars and prices furnished on application.

LEATHER BELTING.

We are agents for the **Jewell Belting Co.'s Superior Leather Belting,** which is unquestionably one of the most reliable Belts made. There are two grades, "Standard" and "Extra," and we keep usually all the leading sizes of each in stock.

WIDTH....	1	1¼	1½	1¾	2	2¼	2½	2¾	3	3½	4	4½	5	5½ in.
Per foot.....	.10	.13	.17	.20	.23	.26	.30	.33	.36	.43	.50	.56	.63	.70
WIDTH.....	6	6½	7	8	9	10	11	12	13	14	15	16	17	18 in.
Price per ft	.76	.83	.90	1.02	1.15	1.29	1.42	1.55	1.68	1.82	1.98	2.14	2.31	2.49
WIDTH.....	19	20	21	22	24	26	28	30	32	34	36	40	44	48 in.
Price per ft	2.66	2.84	3.02	3.20	3.54	3.92	4.30	4.64	5.00	5.35	5.70	6.40	7.10	7.80

Double Belts at Double Prices.

Round Leather Belting.

SIZE.....	1/8	3/16	1/4	5/16	3/8	1/2	5/8	3/4	7/8	1 in.
Solid, per foot..........	.05	.07	.10	.14	.18	.....	...	.....	.....	.....
Twisted, "	.06	.10	.14	.18	.22	.30	.36	.46	.60	.72

Machine Cut Lace.

WIDTH..........	1/4	3/8	1/2	5/8	3/4 in.
Per 100 feet..........	1.00	1.50	2.00	2.75	3.25

Best quality Lace Leather, in the side, 20 cents per square foot.

Improved Pointed Belt Hooks.

NUMBER.	15	14	13	12	11	10	9	8	7	6	5	4 in.
Per thousand.......	2.00	2.40	2.60	2.80	3.00	3.50	4.00	5.00	6.00	8.50	11.00	14.00

Metallic Roofing, Siding, Shingles and Ceiling.

We are Manufacturers' Agents for, and keep constantly in stock, a large assortment of the various kinds of Steel and Iron Roofing, Corrugated Iron Siding, Iron Ceiling, and Metallic Shingles. Circulars and prices furnished on application.

Vulcanized Rubber Packing.

THICKNESS	$\frac{1}{64}$	$\frac{1}{32}$	$\frac{1}{16}$	$\frac{3}{32}$	$\frac{1}{8}$	$\frac{3}{16}$	$\frac{1}{4}$
1-Ply, Per lb.	.70	.65	.60	.55	.55	.55	.55
2-Ply, "	....	....	.63	.58	.55	.55	.55
3-Ply, "	....	....	.66	.61	.58	.55	.55
4 Ply, "	....	....	....	....	.61	.58	.55

One-ply of cloth to every $\frac{1}{16}$ inch thickness.

Three Cents per pound additional will be charged for each extra ply of cloth. Each cloth, whether insertion or on outside, to count as one-ply.

All Cloth Insertion or Plain Packing is one yard wide and any length desired.

Fibrous Gaskets, for Man-hole Plates, Steam Chests, Cylinder Heads, &c., $\frac{1}{8}$ thick, seamless, 90c. per lb., $\frac{5}{32}$ and over.................. $0.80 per lb.

Cloth Insertion Gaskets, Washers, Rings, &c., $\frac{1}{8}$ thick, seamless, $1.25 per lb., $\frac{5}{32}$ and over.................. 1.00

Round Packing, with Duck outside for Stuffing Boxes, Piston Rods, &c., from one-fourth of an inch to two inches diameter.............. .85 "

Square Piston and Valve-Rod Packing, of all sizes, cut to the most exact dimensions, very convenient, no trouble experienced in packing with it, and more durable than any other Packing ever used,.................. .85 "

Special orders for Gaskets, Valves, &c., of any size or pattern that we do not keep on hand can be executed within one week from receipt of order.

Sheet Steam Packing.

Rainbow, per lb.................. .80c.

Jenkins or Usudurin, per lb.................. .80c.

Asbestos Sheating, per lb.................. .15c.

Asbestos Mill Board, per lb.................. .15c.

Wire Insertion Rubber, per lb.................. .50c.

Piston Steam or Water Packing.

Square Flax Packing, $\frac{1}{4}$ inch and larger,		per lb..	$ .85
American Plumbago Packing..................		"	.75
Empire Round Rubber Core Packing........		"	.60
Imperial Oval	"	"	.75
Asbestos Wick	"	"	.50
Italian Hemp	"	"	.25
Soap Stone	"	"	.20
Square Duck	"	"	.85
Crandall or Garlock	"	"	1.20

Rubber Washers for Water Gauges.

Per dozen.........½ inch, $ .40	⅝ inch, $ .50	¾ inch, $ .60

Lead Gaskets for Packing Unions.

SIZE....................	⅜	½	¾	1	1¼	1½	2
Per 100..................................	.55	.60	.65	.70	.85	1.00	1.35

Cotton Waste.

Machine Picked.

Best White, per pound,... .14

Emery Paper—Per Ream.

Nos. 00, 0, ½, 1, 1½, $6.50	No. 2, $7.50	No. 2½, $9.50	No. 3, $11.50

Emery Cloth—Per Ream.

Nos. 00, 0, ½, 1, 1½, $18.00	No. 2, $20.00	No. 2½, $24.00	No. 3, $26.00

Sand Paper.

Nos. 00, 0, ½, 1, 1½, per ream, $4.50	No. 2, 2½ and 3, per ream, $5.00

Galvanized Fire Buckets.

10 Quart, per dozen,....................................	$ 4.50
12 Quart, "	5.00
14 Quart, "	5.50

PIPE COVERING.

Hair Felting.

FOR COVERING STEAM PIPES.

Directions for Applying.

FIRST.—An inside lining next the iron, of one or more thicknesses of Asbestos Sheathing.

SECOND.—A thickness of Hair Felting.

THIRD.—An outside covering of Heavy Canvas.

We furnish all materials for this work. For price of Asbestos Sheathing, see previous pages.

½ inch Hair Felting, per square foot,....................................	$ .06
¾ " " " "	.07
1 " " " "	.08
Canvas Duck, 30 inches wide, per yard,....................................	.15

Patent Removable Non-Conducting Sectional Pipe Coverings.

Tee Closed.

Ell Open.

Cross Open.

Furnished in Sections, 3 feet long. Each section is cut longitudinally on one side, so as to slip on the pipe and the joint afterwards fastened with straps.

Size.	Price per Lineal Foot Canvas, Jacketed	Ells.	Tees.	Globe Valves	Crosses.	Weight per Lineal Foot.
½ inch.	$0 15	$0 16	$0 24	$0 20	$0 28	8 ozs.
¾ "	16	20	26	20	34	9 "
1 "	18	20	26	20	34	10 "
1¼ "	20	20	26	20	34	12 "
1½ "	22	20	26	20	34	15 "
2 "	24	22	29	22	38	18 "
2½ "	27	25	33	33	42	20 "
3 "	30	29	38	38	48	24 "
3½ "	34	32	42	42	52	26 "
4 "	38	35	47	47	60	30 "
4½ "	42	40	52	52	64	38 "
5 "	46	46	60	60	72	44 "
6 "	50	52	72	72	80	48 "
7 "	55	66	96	96	88	55 "
8 "	60	80	1 08	1 08	96	65 "
9 "	65	88	1 20	1 20	1 08	75 "
10 "	75	1 00	1 40	1 40	1 20	85 "

Blocks for Covering Boilers, Drums, Cylinders, Kettles, Heaters, Tanks, &c.

PRICE:—1 in. blocks, 20c. per square ft. 1⅜ in. blocks, 25c. per square ft. Canvas covered blocks 5c. per square ft., extra.

SIZES:—3 in. wide by 18 in. long; 6 in. wide by 36 in. long. Blocks cut to special size and thickness to order. Curved blocks for Locomotive work made to order.

Plastic Covering [dry.] Prepared Carbonate of Magnesia and Fibre for trowel work, per bbl., $5.00.

Circulars and Discounts furnished on application.

STEAM HEATING.

We have had large experience in Heating by Steam all kinds of Buildings, Mills, Factories, Dwellings, Drying Kilns, &c., and will furnish and fit pipe to order, cut to diagram, on short notice.

We have ample machinery for fitting all sizes of pipe into any required shape and can make to order

STEAM PIPE COILS

of every description, and furnish them to the trade at the lowest prices. Following cuts show the principal varieties:

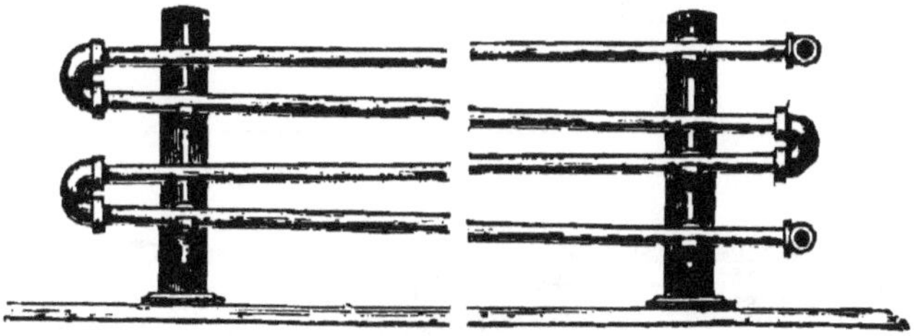

With Return Bends Tapped on Slant.

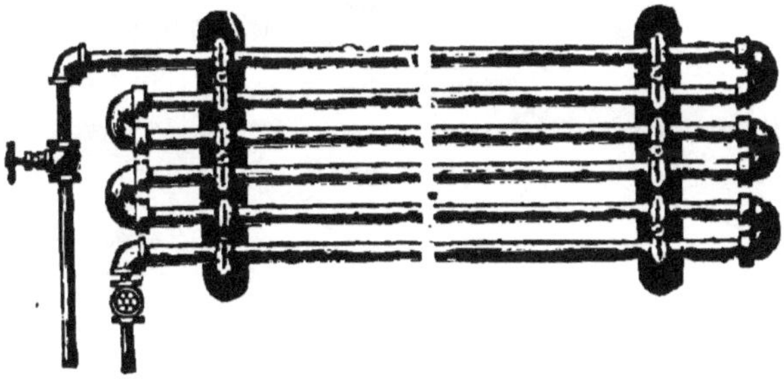

With Return Bends Tapped Straight.

PIPE COILS.—Continued.

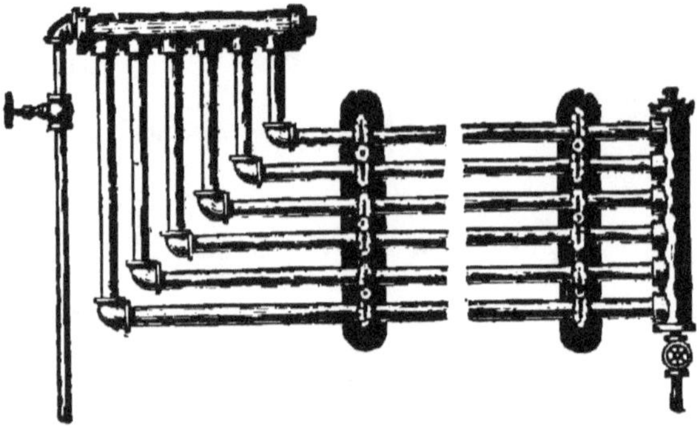

With Branch Tees.—In any Desired Shape.

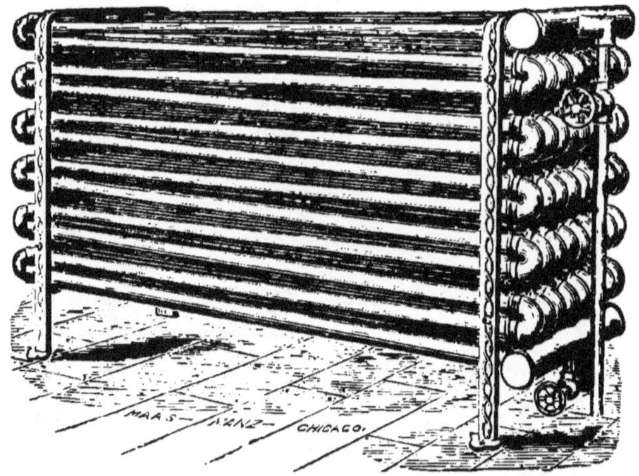

Box Coils of any Size Pipe, any Length, Height or Width.

Finished and proved, ready for use, made to order.

Vertical Tube Radiators.

Painted and ornamented in any style at an extra charge. Furnished as shown above with Open Iron Tops, at the prices stated on next page.

Rectangular.

No. of Rows.	Tubes in each Row.	Surface, Sq ft.	Length. ft. in.	Width. in.	Valves Requir'd Supply. in.	Valves Requir'd Return. in.	Italian Marble Tops, Extra, Net.
Single Row.	1x 4	4	10	4¼	¾	¾	$ .86
	1x 6	6	1— 2	4¼	¾	¾	1.15
	1x 8	8	1— 6	4¼	¾	¾	1.38
	1x12	12	2— 2	4¼	¾	¾	1.90
	1x16	16	2—10	4¼	¾	¾	2.40
	1x20	20	3— 5	4¼	1	¾	2.90
	1x24	24	4— 2	4¼	1	¾	3.30
	1x28	28	4—10	4¼	1	¾	3.75
	1x32	32	5— 6	4¼	1	¾	4.20
	1x38	38	6— 6	4¼	1¼	1	4.90
Two Rows.	2x 4	8	10	6¼	¾	¾	.86
	2x 8	16	1— 6	6¼	¾	¾	1.38
	2x10	20	1—10	6¼	1	¾	1.64
	2x12	24	2— 2	6¼	1	¾	1.90
	2x16	32	2—10	6¼	1	¾	2.40
	2x20	40	3— 6	6¼	1	¾	2.90
	2x24	48	4— 2	6¼	1¼	1	3.40
	2x28	56	4—10	6¼	1¼	1	3.90
	2x32	64	5— 6	6¼	1¼	1	4.40
	2x38	76	6— 6	6¼	1¼	1	5.00
Three Rows.	3 x 4	12	10	8¼	¾	¾	$1.17
	3 x 8	24	1— 6	8¼	¾	¾	1.85
	3 x 12	36	2— 2	8¼	1	¾	2.50
	3 x 16	48	2—10	8¼	1	¾	3.15
	3 x 20	60	3— 6	8¼	1¼	1	3.80
	3 x 24	72	4— 2	8¼	1¼	1	4.40
	3 x 28	84	4—10	8¼	1¼	1	5.10
	3 x 32	96	5— 6	8¼	1¼	1	5.80
	3 x 38	114	6— 6	8¼	1¼	1	7.00
Four Rows.	4 x 4	16	10	10	¾	¾	1.55
	4 x 8	32	1— 6	10	1	¾	2.40
	4 x 12	48	2— 2	10	1	¾	3.20
	4 x 16	64	2—10	10	1¼	1	4.00
	4 x 20	80	3— 6	10	1¼	1	4.85
	4 x 24	96	4— 2	10	1¼	1¼	5.65
	4 x 28	112	4—10	10	1¼	1¼	6.50
	4 x 32	128	5— 6	10	1¼	1¼	7.30

CIRCULAR RADIATORS.

	Surface. Square Feet.	Size.	Valves required.		Italian Marble Tops. Extra—Net.
			Supply.	Return.	
		ft. in.			
No. 1, Circular..	18	Diam. 1— 2	3/4	3/4	$1.40
" 2, " ...	30	" 1— 6	3/4	3/4	2.20
" 3, " ...	54	" 1—11	1	3/4	3.50
" 4, " ...	72	" 2— 2	1	3/4	4.35
" 5, " ...	102	" 2—10	1 1/4	1	7.35
" 6, " ...	130	" 3— 2	1 1/4	1 1/4	9.10
" 7, " ...	160	" 3— 2	1 1/4	1 1/4	9.50
In halves to sur'nd column. " 10, " ...	56	" 2— 2	1 1/4	1 1/4	7.00
" 11, " ...	80	" 2— 4	1 1/4	1 1/4	8.00
" 12, " ...	102	" 2— 9	1 1/4	1 1/4	9.50
" 13, " ...	130	" 3— 2	1 1/4	1 1/4	10.00
" 14, " ...	160	" 3— 2	1 1/4	1 1/4	10.00

Price for Radiators.

Rectangular shape, ...50 cents per tube.
Circular " ....55 " "
" " for Columns... .....60 " "

Radiators may be made of every desired height; the standard height is 35 inches. A Circular Radiator with marble top makes a handsome table.

The dove-colored marble known as the "Knoxville" is well adapted to radiators, as it is not discolored by heat. It costs 25 per cent. more than prices in above table.

CAST IRON RADIATORS.

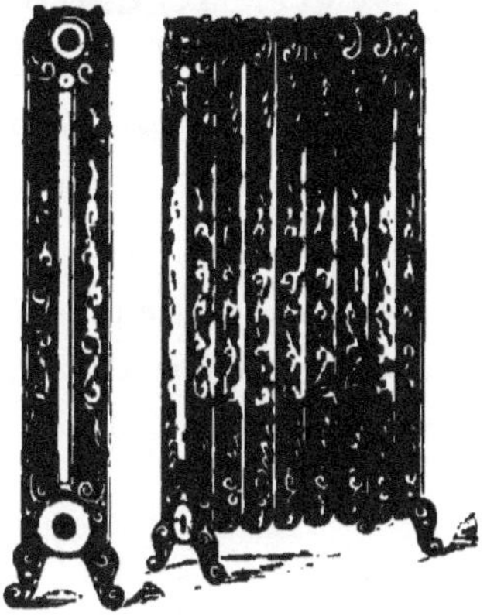

No. of Loops	Length Inches.	Openings for Single Pipe.	Two Pipe Openings. Supply. Return	Heating Surface—Square Feet. 45 Inches High.	38 Inches High.	30 Inches High.	24 Inches High.	18 ins High
2	5.2	1 Inch.	1 x ¾	10	8	6⅔	5⅓	4
3	7.8	1 "	1 ¾	15	12	10	8	6
4	10.4	1 "	1 ¾	20	16	13⅓	10⅔	8
5	13.	1 "	1 ¾	25	20	16⅔	13⅓	10
6	15.6	1¼ "	1 ¾	30	24	20	16	12
7	8.2	1¼ "	1 ¾	35	28	23⅓	18⅔	14
8	20.8	1¼ "	1 ¾	40	32	26⅔	21⅓	16
9	23.4	1¼ "	1 ¾	45	36	30	24	18
10	26.	1¼ "	1 ¾	50	40	33⅓	26⅔	20
11	28.6	1¼ "	1¼ x ¾	55	44	36⅔	29⅓	22
12	31.2	1¼ "	1¼ ¾	60	48	40	32	24
13	33.8	1¼ "	1¼ ¾	65	52	43⅓	34⅔	26
14	36.4	1¼ "	1¼ ¾	70	56	46⅔	37⅓	28
15	39.	1¼ "	1¼ x 1	75	60	50	40	30
16	41.6	1½ "	1¼ 1	80	64	53⅓	42⅔	32
17	44.2	1½ "	1¼ 1	85	68	56⅔	45⅓	34
18	46.8	1½ "	1¼ 1	90	72	60	48	36
19	49.4	1½ "	1¼ 1	95	76	63⅓	50⅔	38
20	52.	1½ "	1¼ 1	100	80	66⅔	53⅓	40
21	54.6	1½ "	1½ x 1	105	84	70	56	42
22	57.2	1½ "	1½ 1	110	88	73⅓	58⅔	44
23	59.8	1½ "	1½ 1	115	92	76⅔	61⅓	46
24	62.4	1½ "	1½ 1	120	96	80	64	48
25	65.	1½ "	1½ 1	125	100	83⅓	66⅔	50
26	67.6	2 "	1½ 1¼	130	104	86⅔	69⅓	52
27	70.2	2 "	1½ 1¼	135	108	90	72	54
28	72.8	2 "	1½ 1¼	140	112	93⅓	74⅔	56
29	75.4	2 "	1½ 1¼	145	116	96⅔	77⅓	58
30	78.	2 "	1½ 1¼	150	120	100	80	60
31	80.6	2 "	1½ 1¼	155	124	103⅓	82⅔	62
32	83.2	2 "	1½ 1¼	160	128	106⅔	85⅓	64

Width of loop, 7¼ inches. | Width across feet, 9¼.

When not ordered otherwise, Radiators will be tapped as above. If openings varying from the above are required, they will be provided without extra charge.

Hot Water Radiators.

The heights and capacities of our Hot Water Radiators are the same as in the Steam Radiators. The flow and return openings are tapped as follows:

Radiators containing 40 square feet and under,........... 1 inch.
Above 40, but not exceeding 72 square feet,............... 1¼ inch.
Above 72 square feet,... 1½ inch.

Height, inches..	18	24	30	38	45
Area per loop, square feet..............................	2	2⅔	3⅓	4	5
Steam per square ft..	30	32	27	24	23½
Hot-Water, per square ft..................................	42	35	30	27	26½

We also furnish when required the **Bundy Cast Radiator**; The **Gold Pin Radiator,** (indirect,) or any other variety at the lowest Manufacturers' Rates.

1 ft. of Heating Surface will heat 50 ft. cubic air, varying in accordance with conditions.

Asbestos Cement.

For Steam Pipes, Boilers, etc...	per barrel,	$3.00
For Hot Blast Pipes, Flues, etc...	"	3.75

(One barrel covers 42 square feet, flat surface, 1 inch thick.)

Mineral Wool.

In about 50 lb. sacks, per lb.. $ .02

GAS FIXTURE FITTINGS.

Two Light Pendant Cocks.

SIZE,	3/8 x 3/8	3/8 x 1/4	3/8 x 1/8	1/4 x 1/8 in.
Per dozen,	9.10	9.10	9.10	8.45

L Pendant Cocks.

SIZE,	[illegible] x [illegible]	3/8 x 1/4	1/4 x 1/8 in.
Per dozen,	5.85	5.20	4.99

L Burner Cocks.

SIZE,	1/2	3/8	1/4	1/8 in.
Per dozen,	6.20	5.20	4.55	4.25

Chandelier Cocks.

SIZE,	1/2 x 1/2	3/8 x 3/8	3/8 x 1/4	1/4 x 1/4	1/4 x 1/8	1/8 x 1/8 in.
Per doz.,	8.00	5.40	5.00	4.50	4.40	3.50

Pillar Cocks.

SIZE,	3/4	1/2	3/8	1/4	1/8 in.
Per dozen,	6.50	5.20	4.55	4.25	3.90

Bracket Swing Cocks.

SIZE,	3/8 x 3/8	3/8 x 1/4	3/8 x 1/8 in.
Per dozen,	9.10	8.45	8.15

Revolving Pendant Cocks.

SIZE,	3/8 x 1/4	3/8 x 1/8	1/4 x 1/4	1/4 x 1/8 in.
Per dozen,	8.15	7.80	7.50	7.15

Street Lamp Cocks.

SIZE,	3/4 x 1/8	1/2 x 1/8	3/8 x 1/8 in.
Per dozen,	6.50	5.85	5.55
With Lever Key, per dozen,	7.80	7.15	6.50

Independent Cocks.

SIZE,	3/8 in.
Per dozen,	6.50

Bracket Bodies.

For Two Brackets.

SIZE,	3/8 x 3/8 in.
Per dozen,	3.90

Pillar Bodies.

Two Light.

SIZE,	3/8 x 3/8 in.
Per dozen,	5.20

Universal Swings.

SIZE,	3/8 x 1/4	1/4 x 1/4	1/4 x 1/8	1/8 x 1/8 in.
Per dozen,	8.80	8.45	8.15	7.80

Universal Top Swings.

SISE,	3/8 x 3/8	3/8 x 1/4 in.
Per dozen,	6.20	5.55

Universal Bracket Cocks.

SIZE,	3/8 x 3/8	3/8 x 1/4	3/8 x 1/8 in.
Per dozen,	13.00	12.35	12.05

Universal Pendant Cocks.

SIZE,	1/4 x 1/8 in.
Per dozen,	9.40

Top Swings.

SIZE,	3/8 x 3/8	3/8 x 1/4	3/8 x 1/8 in.
Male or Female, per dozen,	6.20	5.55	5.20

Middle Swings.

SIZE,	1/4 x 1/4	1/4 x 1/8	1/8 x 1/8 in.
Per dozen,	4.90	4.55	4.25

Side Nozzles.

SIZE,	3/8	1/4	1/8 in.
Per dozen,	2.30	1.65	1.00

Straight Nozzles.

SIZE,	3/8	1/4	1/8 in.
Per dozen,	1.95	1.65	1.00

Stiff Joints.

SIZE,	1/2 x 1/2	1/2 x 3/8	1/2 x 1/4	3/8 x 3/8	3/8 x 1/4	3/8 x 1/8 in.
Per dozen,	3.25	2.60	2.60	1.95	1.85	1.55

STAPLE GAS FIXTURES, &c.

"L" Pendant.

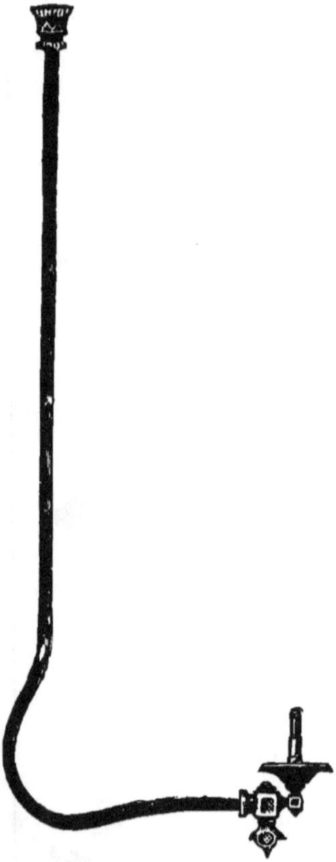

30 in., Polished, Brass, Each, $1.50
30 " Iron, Bronzed, " 1.25

Hall Light.

Polished—Brass.

Each, $2.30.

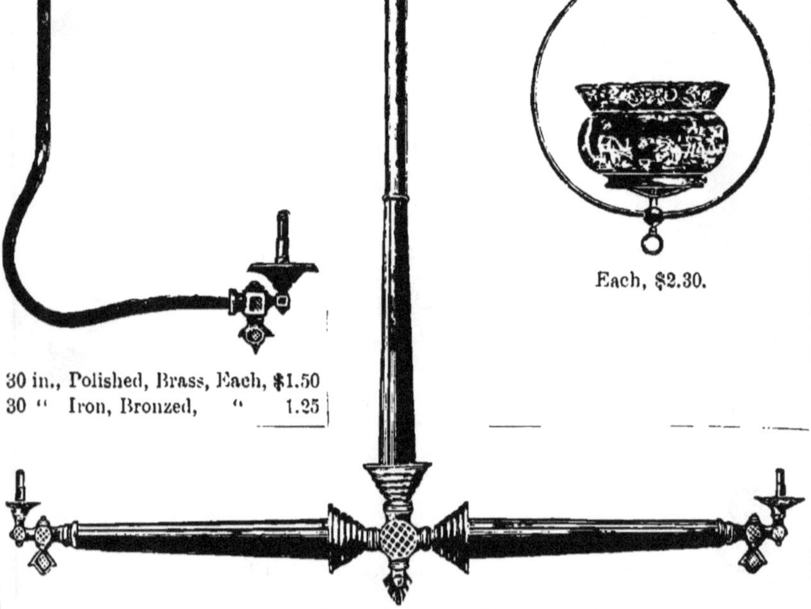

Sleeve Pendant.—Bronzed.

Light.	Stem.	Spread.	Price.	Light.	Stem.	Spread.	Price.
2	24 in.	24 in.	$2.80	2	42 in.	42 in.	$3.70
2	30 "	30 "	3.10	2	42 "	48 "	4.00
2	36 "	36 "	3.38				

Stiff Bracket.

Brass, Gilt.

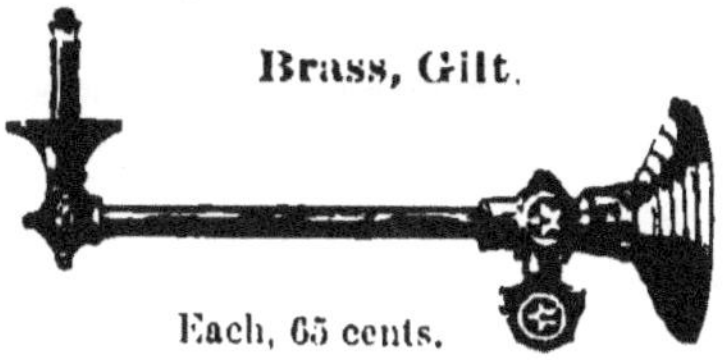

Each, 65 cents.

One Swing Bracket.

ROUND TUBE.

Brass, Gilt.

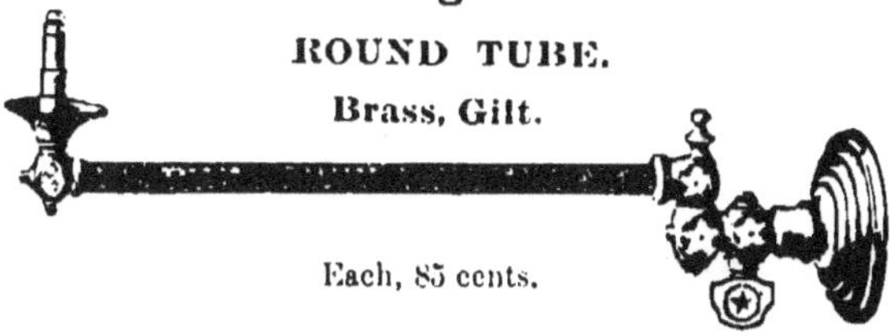

Each, 85 cents.

Two Swing Bracket.

ROUND TUBE. Brass, Gilt.

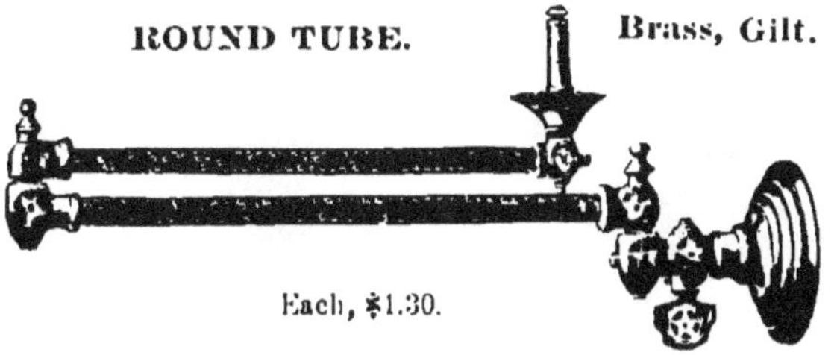

Each, $1.30.

Three Swing Bracket.

ROUND TUBE.

Brass, Gilt.

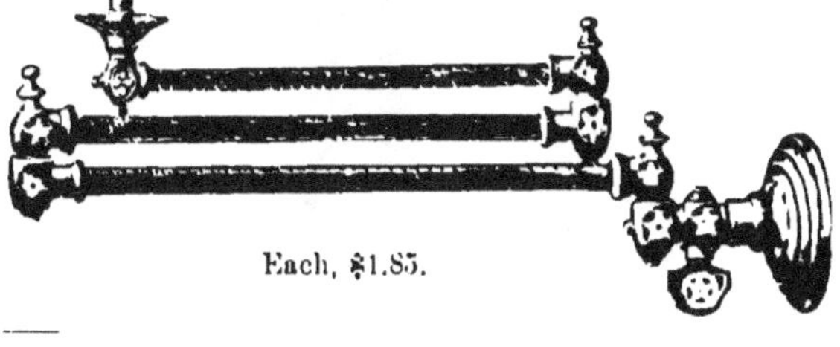

Each, $1.85.

Three Swing Bracket.

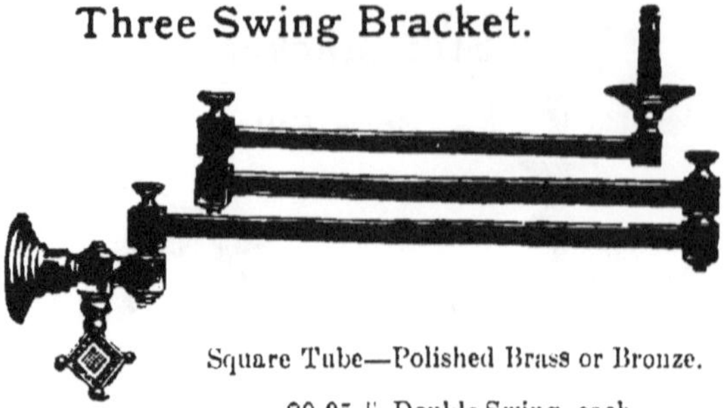

Square Tube—Polished Brass or Bronze.

Stiff, each	$0.95	Double Swing, each	$2.10
Single Swing, each	1.50	Three " "	2.75

GAS BURNERS.

ARGAND.

Per Dozen, $7.00.

EMPIRE.

BASE WITH ADJUSTABLE SCREW CHECK.

Per Gross, including Tips, $12.00

Common Iron.

Per Gross, $7.00

Common Brass.

Per Gross, $5.00.

Brass Lava Tip.

Per Gross, $5.00.

Clough Gasoline.

Per Dozen, $5.50.

MATCHLESS SELF-LIGHTING BURNERS.

No. 3 B, Per Dozen, $8.50.

BRASS PILLARS.
For Lava Tips.

Per Gross,........ 2.20

Iron
Scotch Tip.

Per Gross, 3 50

Lava Tip.

Per Gross, 1.80

Burnor Cleaner.

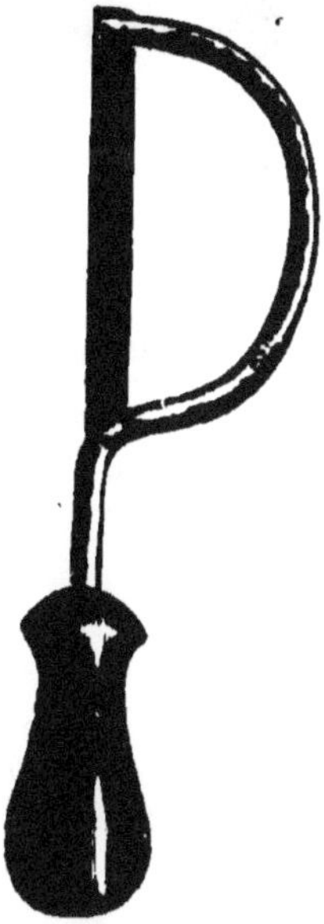

Per dozen............. 1.20

BOX PLATE.
Polish'd Brass or Bronze

Per dozen,........................ 1.50

BRACKET BACK.
Stamped Brass.

$2\frac{1}{2}$ in. per dozen,........................... 50c.

BURNER CUP.
Brass.

Per Gross,............. .... 2.50

BURNER CUP.
Brass or Bronze.

Per Gross,........................ 3.00

BRACKET BACK.
Spun.
Brass or Bronze.

$2\frac{1}{2}$ in.	per dozen,.........	.50
3 "	"	.70
4 "	"	1.50

Gas Torch and Key.

No. 620, 30 inches long, per dozen, ... 24.00
No. 300, 30 " " " " 15.00

Gas Key.

No. 410, 25 inches long, per dozen, ... 8.00
No. 475, 27 " " " " 10.00

Taper Slides or Holders.

No. 422, 24 inches long, per dozen, ... 3.00
No. 423, 25 " " " " 4.00

Wax Tapers.

Howchin's Patent Paragon, with braided wick and fringed ends, 18 inches long.

No. 90, 120 Tapers in box, per dozen boxes 9.60
No. 91, 60 " " " " 4.80
No. 92, 30 " " " " 2.40

Mica Smoke Canopies.

No. 423, 6½ inch, per dozen, ...$3.50

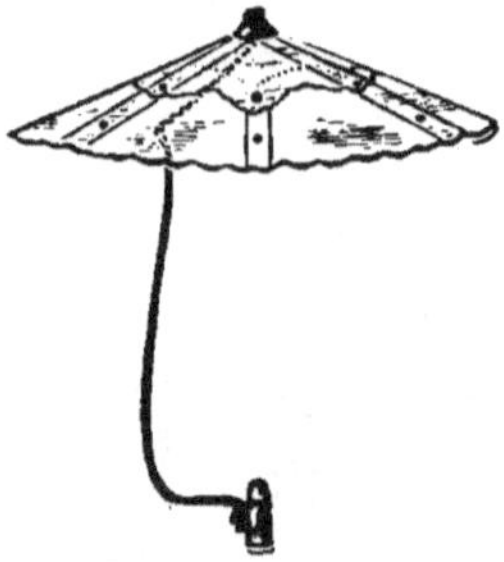

No. 520, 6½ inch, per dozen, ...$3.50

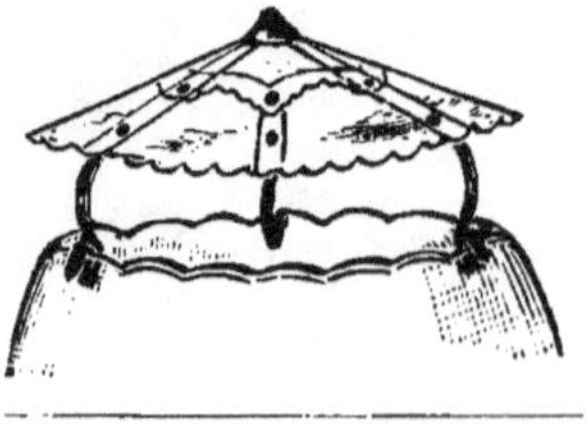

No. 222, 6½ inch, per dozen, ...$3.20

No. 224, per dozen, ...$1.50

Lengthening Pieces.

Size,	3/8 x 3/8	3/8 x 1/4 in.
Per dozen,	1.95	1.95

Ball Joints.

Size,	3/8 x 3/8	1/2 x 3/8	1/2 x 1/2	3/4 x 1/2	1 x 3/4 in.
Per dozen,	81.90	44.20	44.20	81.90	117.00

Chandelier Hooks and Staples.

Size,	3/8	1/2 in.
Brass Hooks, per dozen,	2.60	2.95
Iron Staples, "	1.65	1.65

Miscellaneous Gas Trimmings.

Argand Chimneys, 6 inches long, cut ends	Per dozen,	.80
Polished ends	"	1.00
Opal Squat Globes, 7½ x 5	"	4.50
Ring Top Cone Opal Shades for Argand Burners, 10 inch	"	5.50
Argand Spring Holder, 10 inch	"	2.00
Argand Extension Holders, 10 inch	"	4.00
Fancy Globe Holders, Brass or Bronzed, 4 inch	"	1.00
" " " " " 5 inch	"	1.20
Stork Necks for Portables	"	2.50
Drop Light Sockets, 5/16 and 3/8	"	3.00
Wire Goose Necks for Portables	"	2.00
Mohair Tube for Portables in six foot lengths	Per length,	1.50
Tan Tubing for Gas Stoves, 5/16	Per foot,	.24
Cloth Covered Tubing with Patent ends, 4, 6 and 8 feet lengths	"	.16

Chandeliers and Fancy Brackets, (photographs of which will be loaned on application) furnished at manufacturers' prices, and shipped direct from factory.

TOOLS.

Pipe Threading Machines.

No. 1, Hand Machine, cuts and threads, ¼ to 2 inch pipe, inclusive.........$ 90.00
No. 1, Power " " " ¼ to 2 inch " " 110.00

Forbes Pipe Machine.

These Machines have Opening and Adjustable Dies, Self-Centering Vises and Cut-off Attachment.

No.	Range.	Weight.	Each.
2	2½ to 4 inches. R. H.	175 lbs.	$ 85.00
2½	1½ to 4 inches. R. H.	175 lbs.	100.00
2½ A	1½ to 4 inches. R. & L.	180 lbs.	115.00
2½ B	1 to 4 inches. R. H.	180 lbs.	110.00
2½ C	1 to 4 inches. R. & L.	185 lbs.	130.00
3	4 to 6 inches. R. H.	310 lbs.	115.00
3 A	3½ to 6 inches. R. H.	315 lbs.	130.00
3 B	2½ to 5 inches. R. H.	320 lbs.	150.00
3 C	2½ to 6 inches. R. H.	325 lbs.	175.00
3 D	2½ to 8 inches. R. H.	600 lbs.	325.00

Circulars and Prices on other sizes furnished on application.

Malleable Stocks with Solid Dies.

No.	Range.	Dimension of Dies.	Die Plate Complete.	Without Dies.	Extra Dies.	Extra Guides.
1½	¼ to 1	2½ x 2½ x ¾	15.00	5.00	2.00	.35
2	¾ to 1¼	3 x 3 x ¾	13.50	6.00	2.50	.45
3	1¼ to 2	4 x 4 x ⅞	20.00	9.50	3.50	.60
4	2½ to 3	5 x 5 x 1¼	43.00	25.00	9 00	1.00

Nos. 3 and 4 with leader screws.

Armstrong's Adjustable Stocks and Dies.

No.	Range.	Complete.	Stock Only.	Extra Dies.	Extra Guides.
1	1/8 to 1/2	9.00	3.00	1.20	.20
2	1/4 to 1	12.00	3.50	1.50	.25
2½	½ to 1¼	12.00	4.50	3.00	.40
3	1¼ to 2	20.00	7.00	4.00	.50

Nos. 1 and 2 Packed in Cabinet Cases.
No. 2½ is fitted with Double Ended Dies.
We also keep in Stock Hart's Duplex Die Stock. Prices furnished on application.

Ratchet Stocks and Dies, Miller's.

Malleable, with Solid Dies.

No.	Threads.	Dimensions of Dies.	Complete.	Without Dies.	Extra Dies.	Extra Guides.
1½	¼ to 1 in.	2½x2½x ¾	15.00	7.50	1.50	.25
2	¾ to 1¼ "	3 x3 x ¾	18.50	13.00	1.80	.35
3	1¼ to 2 "	4 x4 x 7/8	20.00	12.50	2.50	.45
4	2½ to 3 "	5 x5 x1¼	43.00	29.00	7.00	.75

Nos. 2, 3 and 4 have Leader Screws.
We also furnish Clark's Ratchet Stocks and Dies. Prices on same will be given on application.

Maule's Skeleton Die.

Right or Left Hand.

Size Frame, 2½ in. square × ¾ in. thick, for ½, ¾, 1 in. Pipe, each..........$0.55
" " 4 " " ×1 " " " 1¼, 1½, 2 in. Pipe, each........ 1.00
Special Dies for "Running" or Longscrew threads, 2½ in. × ¾ in., each...... .70
" " " " " " " 4 in. × 1 in., each...... 1.25

Gas Pipe Taps.—Right or Left.

SIZE	1/8	1/4	3/8	1/2	3/4	1	1¼	1½	2	2½	3 in.
Each	1.12	1.25	1.50	1.87	2.50	3.12	3.75	4.62	6 25	10.50	15.00
Threads per in.	27	18	18	14	14	11½	11½	11½	11½	8	8

Gas Pipe Reamers and Drills.

SIZE	1/8	1/4	3/8	1/2	3/4	1	1¼	1½	2	2½	3 in.
Reamers	1.12	1.25	1.50	1.87	2.50	3.12	3.75	4.62	6.25	10.50	15.00
Drills	.30	.30	.30	.35	.45	.60	.75	1.00	1.25	....	....

Humphrey's Combined Drill, Reamer and Tap.

SIZE	½	¾	1	1¼	1½	2
Each	3.00	4.50	6.00	7.25	8.50	10.75

Eureka or Stanwood Pipe Cutters.

No. 1, Cuts from 1/8 inch to 3/4 inch pipe $1.50
No. 2, " 3/4 " 2 " 2.25
No. 3, " 2 " 3 " 7.00

Saunders' Wheel Pipe Cutters.

No. 1, Cuts 1/8 inch to 1 inch pipe .... $ 3 00
No. 2, " 1 " 2 " 4.50
No. 3, " 2 " 3 " 14.00

Comstock Pipe Cutters.

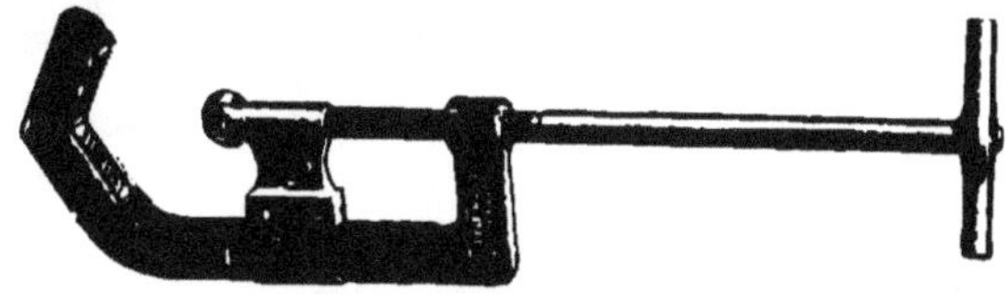

With Attachments for Removing Burr and Scale while Cutting.

No. 1, Cuts from 1/8 to 1 1/4 inches	$ 4.50
No. 2, " " 1 " 2 1/2 "	6.00
No. 3, " " 1 1/2 " 4 "	10.00

Barnes' Three Wheel Pipe Cutters.

No. 1, Cuts 1/8 to 1 inch pipe	$ 4.50
No. 2, " 1/2 " 2 " "	6.00
No. 3, " 1 1/2 " 3 " "	10.00

Armstrong Three Wheel Cutters.

No. 1, Cuts 1/8 to 1 inch pipe	$ 4.50
No. 2, " 1/2 " 2 1/2 " "	6.00
No. 3 " 1 1/2 " 4 " "	20.00

Wheels for Pipe Cutters.

No. 1 .... $0.24 | No. 2 .... $0.32 | No. 3 .... $0.60

Malleable Iron Pipe Vises.

With Hinged or Solid Frame.

No. 1, holds 1/8 to 2 inch, inclusive	$ 8.00
No. 2, " 1 " 3 " "	12.00

Steel Pipe Vises.

Light and convenient to carry. Holds 1/8 to 2 inch pipe $6.00

Combination Vises.

No. 1, holds 1/8 to 2 inch pipe.............................$16.00
No. 2, " 1/4 " 3 " 20.00

Common Pipe Tongs.

Size	1/8	1/4	3/8	1/2	3/4	1	1 1/4	1 1/2	2	2 1/2 in.
Each	.60	.65	.70	.75	.90	1.10	1.30	1.50	1.90	3.50

Brown's Patent Adjustable Tongs.

We also furnish all other vises of different manufacturers' make, at their prices.

Number	1	1 1/2	2	3	4	5	6
Each	1.30	1.65	2.00	3.00	6.00	11.00	25.00
Taking Pipe from	1/8 to 3/4	3/8 to 1	1/2 to 1 1/4	1 to 2	1 1/4 to 3	2 1/2 to 4	3 to 5

Jarecki's Adjustable Pipe Tongs.

Numbers	1	2	3	4	5
Each	3.50	4.00	5.00	9.00	16.00
Taking Pipe from	1/8 to 1	1/4 to 1 1/4	1/2 to 2 1/2	3/4 to 3 1/4	2 1/4 to 6 in.

Robbin's Patent Chain Wrench.

	Length of Lever.	Size of Lever near Claw	Size of Chain	Weight.	Size of Pipe adapted to.	Each.
No. 2	27 in.	1 1/8 in.	5/16 in.	7 lbs.	1 to 2 in.	5.50
" 3	3 feet.	1 1/4 "	5/16 "	12 "	1 1/4 to 4 "	6.25
" 4	4 "	1 1/2 "	3/8 "	24 "	2 to 6 "	9.00
" 5	5 "	1 3/4 "	1/2 "	33 "	2 1/2 to 8 "	12.50
" 6	6 "	2 1/2 "	5/8 "	50 "	4 to 10 "	16.00

Brock's Patent Drop Forged Chain Pipe Wrench.

SIZE........	No. 1.	No. 2.	No. 3.	No. 4.	No. 5.
Each............	3.50	5.50	7.50	11.00	18.00
Capacity........	⅛ to 1½ in	¼ to 2½ in.	¾ to 3½ in.	1½ to 8 in.	2 to 14 in.
Length..........	20 in.	27 in.	37 in.	50 in.	64 in.
Weight..	4¼ lbs.	8 lbs.	15 lbs.	27 lbs.	46 lbs.

Hilts' Chain Pipe Wrench.

Which is **Adjustable,** taking up all slack in the chain, and goes to work immediately. There is no valuable time lost in grasping the pipe

SIZE............	No. 2.	No. 3.	No. 4.	No. 5.	No. 6.	No. 7.
Each	5.50	7.50	9 50	11.00	18.00	30.00
Capacity.............	[illegible] to 2½ in.	1¼ to 4 in.	1½ to 6 in.	2 to 8 in.	2½ to 10 in.	4 to 16 in
Length of Lever...	26 in.	36 in.	46 in.	58 in.	70 in.	80 in.
Weight................	9 lbs.	15 lbs.	25 lbs.	36 lbs.	50 lbs.	85 lbs.
Ext. Chain, Each..	.65	.80	.95	1.25	2.00	3.00
Extra Jaws.........	2.00	3.00	3.50	4.50	5.00	5.50

Trimo Pipe Wrenches.

Length open, in in	6	8	10	14	18	24	36	48
Takes from.......	⅛ in. wire to ⅝ in. pipe.	¼ in. wire to ¾ in. pipe	¼ in. wire to 1 in. pipe.	¼ in. wire to 1½ in. pipe	⅜ in. wire to 2 in. pipe.	¼ in. wire to 2½ in. pipe.	½ in. pipe to 3½ in. pipe.	1 in. pipe to 5 inch pipe.
Each...............	2.00	2.00	2.25	3.00	4.00	6.00	12.00	18.00
Jaw	.67	.67	.75	1.00	1.33	2.00	4.00	6.00
Nut......	.20	.20	.27	.35	.42	.50	.65	.80
Inserted Jaw.....	.25	.25	.33	.50	.55	.65	1.00	1.25
Saddle.............	.20	.20	.27	.35	.42	.50	.65	.80
Rocker.....	.20	.20	.27	.35	.42	.50	.65	.80

In ordering parts state the size of Wrench.

Stillson Wrench.

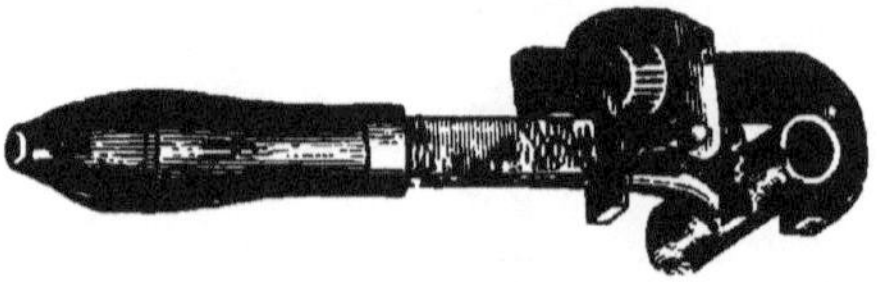

Length open, in in.	6	8	10	14	18	24	36	48
Takes from........	⅛ in. wire to ½ in. pipe.	⅛ in. wire to ¾ in. pipe.	⅛ in. wire to 1 in. pipe.	¼ in. wire to 1¼ in. pipe.	¼ in. wire to 2 in. pipe.	¼ in. wire to 2½ in. pipe.	½ in. pipe to 3½ in. pipe.	1 in. pipe to 5 in. pipe.
Each............	2.00	2.00	2.25	3.00	4.00	6 00	12.00	18.00

The Donohue Combination Wrench.

Length.	10	12	15 in.
Takes pipe from................................	½ to 1	½ to 1½	½ to 2
Short Sleeve....................................	2.00	2.25	3.25
Long Sleeve.....................................	2.25	2.50	3.50

Adjustable Alligator Wrench.

Numbers..................................	7	9	13	15
Holds Pipe..................................	⅛ to ⅜	⅛ to ¾	¼ to 1½	⅜ to 1¼
Length. inches..............................	5¾	10	16	22
Each..	1.50	1.75	2.50	3.00

Screw Wrenches.

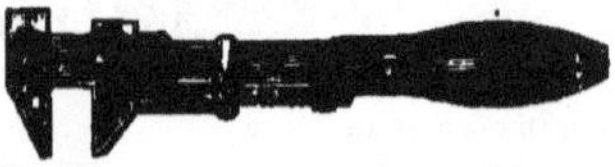

SIZE...........	6	8	10	12	15	18	21 in.
Black, per doz...	9.00	10.00	12.00	14.00	24.00	30.00	36.00
Bright, "	10.00	12.00	14.00	16.00	26.00	32.00	38.00

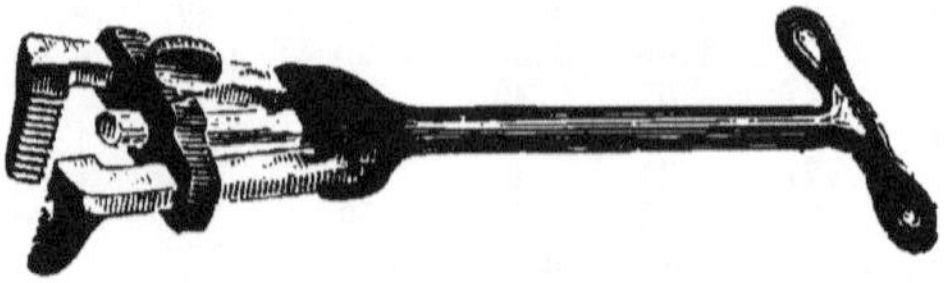

Basin Wrenches.

BUZZELL'S PATENT.

Per doz.......................$22.00

Gas Plyers.

SIZE	5	6	7	8	9	10	12	14 in.
Polished, per dozen	......	......	7.40	8.25	9.25	10.70	13.00	17.00
Burner Plyers, Polished, per doz.	4.80	6.50	10.00	......	......	......	......	......

Gas Fitters' Proving Pumps and Gauges.

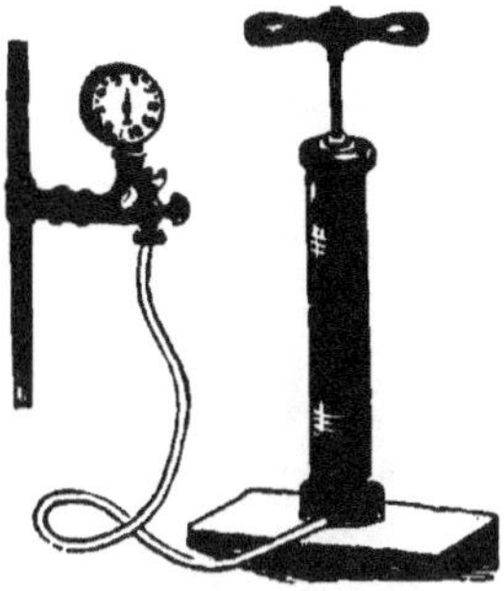

Proving Pump, with Gauge, complete	$25.00
" " and Hose only	15.00
Mercury Gauge comp., with Cock & Ether Cup,	10.00
Proving Pumps only	13.00
Cock and Ether Cup	5.00
Ether Cup only	.50
Mercury Cup only	4.00
Brass Guard for Glass	1.50
Glass only	1.00
Hose only	2.00

Torches, or Gas Fitters' Lamps.

Tin, each	$0.80
Brass, "	1.50

Brass Blow Pipes.

No. 1 .... $ 25 | No. 2 .... $ .35

Breast Drills.

Single Geared	$3.00
Double Geared	4.00

Moore's Patent Ratchet Wrenches.

No. 1,	8 inch Lever	$3.00
No. 2,	10 "	4.00
No. 3,	15 "	5.00
No. 4,	18 "	7.00

Moore's Sleeve Ratchet.

TRIPLE ACTION.

NUMBER	1	2	3	4
Length of Handle	10	12	15	18 in.
Each	6.00	8.00	10.00	12.00

Charcoal Furnaces.

Number	2	3	4	5
Diameter on Top	12	13	14	15 in.
Each	1.50	1.75	2.25	2.50

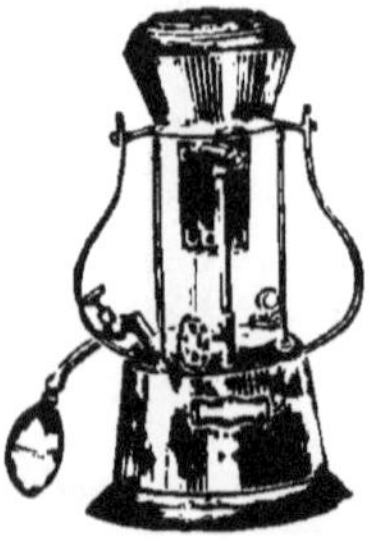

Plumbers' Melting Furnaces.

For Using Gasoline or Naphtha.

Each	$ 9.00

Plumbers' Gasoline Torch.

Each	$ 6.00

Ladles.

SIZE	2½	3	4	5	6	7	8 in.
Each	.25	.30	.40	.55	.70	2.00	2.50

Solder Pots.

Number	1	2	4	5	6	7
Diameter on Top	5	6	8	9	10	12
Each	.50	.70	1.20	1.35	1.80	3.20

Soldering Coppers.

Pointed, per lb	$ .30	Hatchet, per lb	$ .30

Round Irons.

Number	1	2	3
Each	.70	.85	1.00

Files and Rasps.

Size	10	12	14 in.
Half-Round Files, each	.50	.65	.90
Flat " "	.40	.60	.80
Rasps, each	.75	1.05	1.40

Chipping Knife.

Size	4	5	6 in.
Each	.40	.45	.50

Wiping Cloths.

Ticking, Each	$ .15
Moleskin "	.20

Shave Hooks.

Oval, Half-Oval, or Triangle .... Each, $0.40

Bending Pins.—Steel.

Light	$0.25
Heavy	.40

Floor Chisels.

Size	1½	2	2½	3	3½	4 in.
Per dozen	6.00	8.00	11.00	12.00	14.00	16.00

Chisels.

Cold Chisel each........................$0.35 | Calking Chisel, each...................... $0.30

Yarning Chisel, each............ .$0.30

Tap Borders.

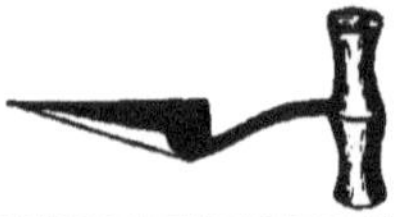

Price, each...$0.25

Hammers.

Plumbers' Hammers, Each..................Small, $ .75 | Large. $ .85

WEIGHT..............................	12 oz.	1 lb.	1¼ lb.	1½ lb	2¼ lb.
No. 2, Machinist, Straight Pein, each	1 00	1.00	1.10	1.20	1.50
No. 1, " Ball " "	1 00	1.00	1.10	1 20	1.50

Screw Drivers.

SIZE...	5 x 6	8 x 10	12 x 14 in.
Each..	.20	.25	.35

Plain Compass.

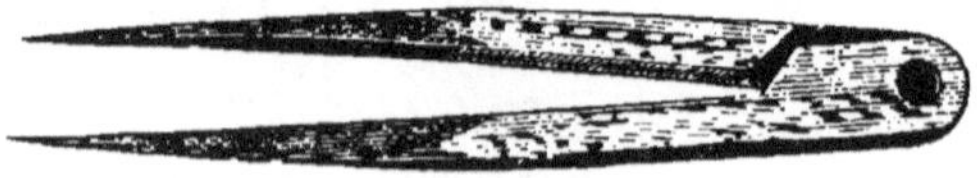

SIZE..............................	5	6	7	8	9	10	12 in.
Per dozen.	3.50	4.00	4.75	5 50	6.50	7.50	8.00

Wing Dividers.

SIZE..............................	5	6	7	8	9	10	12 in.
Per dozen..............................	5.50	5.50	6.50	7.50	9.00	10.00	12.00

Compass Saws.

9, 10 and 12 inch, per dozen .... $ 5.00

Hack Saws.

8 inch, per dozen .... $ 8.00
Extra Blates, per dozen 1.00

Gauge Glass Cutters.

Hunt & Connell's.

Handy.

With Scale to 5 in .... each, $2.00 | Each $0.50

Turn Pins—Boxwood.

NUMBER	1	2	3
Price each	.15	.20	.25

Boxwood Drift Plugs.

SIZE	½	¾	1	1¼	1½	2 in.
Price, each	.05	.05	.10	.15	.20	.25

Dressers.

Hickory, each $0.45
Boxwood, " 1.00

Bossing Sticks.

Dogwood, each .... $0.35 | Boxwood, each $0.75

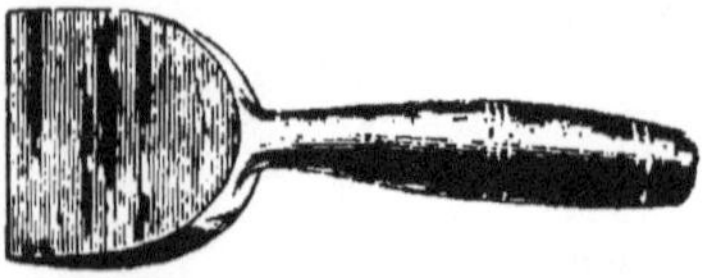

Side Edges.

Hickory, each.......................$0.35
Boxwood, " .... .50

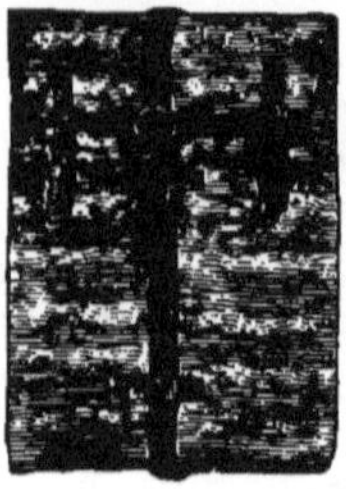

Plumbers' Bags.

Plain, each.....................................$4.50
Leather Bottom, each 5.50
Leather Bottom and Sides, each................... 7.00

Heavy Brass Grease Boxes.

No. 1, for Grease or Rosin, each..............................$1.00
No. 2, " Rosin and Flour, Spun Brass, each.................... 1.50

Tack Moulds.

Plain, Single..each, $1.50
" Double .... " 2.00
Fancy, Single.. " 1.50
" Double.. " 2.00

Springs for Bending Lead Pipe.

SIZE	1 or 1¼	1½	2 in.
Each..........	1.50	1.50	2.00

Set of three above Sizes, $5 00.

CHARLES MILLAR & SON'S

Scale of Weights of Lead Pipes.

Caliber and Mark.			Weight per foot. lbs.	oz.	Caliber and Mark.			Weight per foot. lbs.	oz.	Caliber and Mark.			Weight per foot. lbs.	oz.
1/4	inch	E	0	5	3/4	inch,	C	1	12	2	inch,	AAA	10	11
3/8	"	AAA	1	12	3/4	"	D	1	3	2	"	AA	8	14
3/8	"	AA	1	5	3/4	"	E	1	0	2	"	A	7	0
3/8	"	A	1	2	1	"	AAA	6	0	2	"	B	6	0
3/8	"	B	1	0	1	"	AA	4	8	2	"	C	5	0
3/8	"	C	0	14	1	"	A	4	0	2	"	D	4	0
3/8	"		0	10	1	"	B	3	4	2	"	Waste	3	0
3/8	"	D	0	7	1	"	C	2	8	2½	"	3/8 thick	16	11
7/16	"		0	9½	1	"	D	2	4	2½	"	5/16 "	13	10
1/2	"	AAA	3	0	1	"	D	2	0	2½	"	1/4 "	10	10
1/2	"		2	8	1	"	E	1	8	2½	"	3/16 "	7	13
1/2	"	AA	2	0	1¼	"	AAA	6	12	2½	"	Waste	6	0
1/2	"	A	1	10	1¼	"	AA	5	12	2½	"	"	4	0
1/2	"	B	1	3	1¼	"	A	4	11	3	"	3/8 thick	19	9
1/2	"	C	1	0	1¼	"	B	3	11	3	"	5/16 "	16	0
1/2	"	D	0	12	1¼	"	C	3	0	3	"	1/4 "	12	9
1/2	"	D	0	10	1¼	"	D	2	8	3	"	3/16 "	9	4
1/2	"	D	0	9	1¼	"	E	2	0	3	"	Waste	5	0
5/8	"	AAA	3	8	1½	"	AAA	8	0	3	"	"	3	8
5/8	"	AA	2	12	1½	"	AA	7	0	3½	"	3/8 thick	22	8
5/8	"	A	2	8	1½	"	A	6	4	3½	"	5/16 "	18	7
5/8	"	B	2	0	1½	"	B	5	0	3½	"	1/4 "	14	8
5/8	"	C	1	7	1½	"	C	4	4	3½	"	3/16 "	10	12
5/8	"	D	1	4	1½	"	D	3	8	4	"	3/8 "	25	6
5/8	"	D	1	0	1½	"	D	3	0	4	"	5/16 "	20	14
5/8	"	E	0	12	1½	"	Waste, E	2	0	4	"	1/4 "	16	7
3/4	"	AAA	4	14	1¾	"	AA	8	8	4	"	3/16 "	12	2
3/4	"		4	0	1¾	"	A	6	7	4	"	Waste	8	0
3/4	"	AA	3	8	1¾	"	B	5	0	4	"	"	6	0
3/4	"	A	3	0	1¾	"	C	4	9	4	"	"	5	0
3/4	"	B	2	3	1¾	"	D	3	10	5	"	"	7	8

Lead Tubing.

1/8 inch,	Weight, 3 oz.	1/4 inch,	Weight, 5 oz.

Pure Block Tin Pipe.

Scale of Weights.

Caliber.	Weight per foot.	Weight per foot.	Caliber.	Weight per foot.	Weight per foot.
1/8 inch.	3 oz.	......	5/8 inch	9 oz.	14 oz.
1/4 "	5 "	......	3/4 "	11 "	1 lb.
5/16 "	6½ "	......	1 "	14 "	1 " 4 "
3/8 "	4½ oz. 5 "	8 oz.	1¼ "	......	1 " 8 "
1/2 "	6½ "	10 "			

Tin Lined Lead Pipe.

Weights and Sizes Same as Lead Pipe.

N. B.—Our Machinery is of the latest and most improved description. Our Pipe is made in the most accurate and true manner, from the Best Quality of Pig Metal. NO OLD METALS USED. Great care is taken in Packing and shipping, while our Improved Reels, having large centers, are especially liked by the trade. A trial order is solicited. We desire also to state here, we are not members of "the combination" and probably never will be.

Weights of Sheet Lead.

Wire Gauge,	No. 18.—	Weight per	square foot,	2½	lbs.
"	No. 17,—	"	"	3	"
"	No. 16,—	"	"	3½	"
"	No. 15.—	"	"	4	"
"	No. 14.—	"	"	4½	"
"	No. 13,—	"	"	5	"
"	No. 12,—	"	"	6	"
"	No. 11,—	"	"	7	"
"	No. 10,—	"	"	8	"

Pig Lead, Block Tin, Spelter, Antimony, &c.,

Always on hand, and supplied to the trade at market rates.

Babbitt Metal.

Nos	4	4a	3	2	1	Extra.	Genuine.
Per lb.....................................	7½	9	11	16	19	30	35c.

SOLDER.

We are making from strictly new material, solders of all kinds. We have been obliged to do this on account of the unreliability of the Solders generally sold, which are made from old materials. We make three grades for tinners' use largely :

CHAS. MILLAR & SON'S BEST, made strictly from equal parts of the best Pig Tin and Lead ;

CHAS. MILLAR & SON'S O. K., made from 45 parts Tin to 55 parts Lead;

CHAS. MILLAR & SON'S NO. 1, made from 40 parts Tin to 60 parts Lead.

Each grade made in Triangular Bars for Canning Factories.

We also make an extra quality of

Plumbers' Wiping Solder.

Our Bar.

This Solder is cast in our special moulds into bars of the above shape. Each bar weighs about five pounds and thus each section of the bar will weigh about one pound. This makes the Solder very convenient for use, many times saving the trouble of weighing it.

Murray's Compressed Lead Sash Weights.

Round or Square.

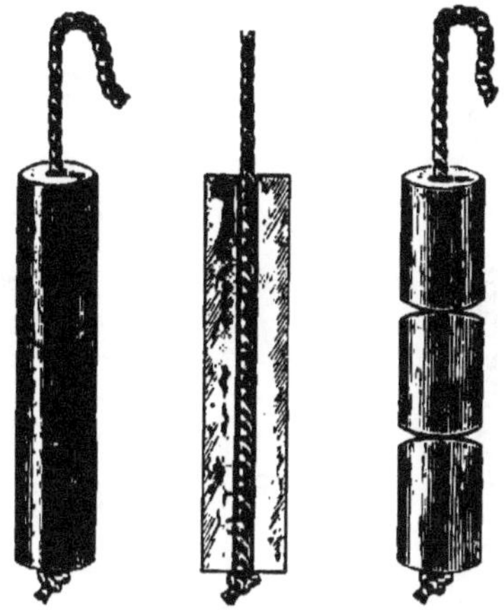

Made from Solid Lead under Hydraulic Pressure, with ⅜ in. hole in center to accommodate the sash cord, securing great density and perfect smoothness. No iron is used in their construction, so that the minimum of space is required. This is frequently important with heavy sash, *as iron weights occupy nearly twice as much space as these.*

They are center balanced, can be cut in sections to accommodate small pockets, have no bolts to unscrew, no eyes to pull out, while the cord tied with knot at lower end, as shown, is less liable to wear. *Their cost is very reasonable* and they are positively in every respect the most convenient and desirable to use.

We make all sizes and can furnish them of exact weight for each sash to order, or in lengths of about ten feet to be sawed up as desired, by the carpenter, when hanging the sash.

Table of Approximate Weights.

Round. Outside Diameter, inches..........	1¼	1⅜	1½	1⅝	1¾	2	2¼	2½
Weight per Lineal Foot, pounds...........	5¼	6⅝	8	9¾	11⅝	15	19	23

N. B.—Nothing we make has been received with so much favor as these Sash Weights, which we have shipped to all sections of this country.

For Square Weights add 25%.

Utica Pipe Foundry Co.'s Dimensions and Weights of Cast Iron Water Pipe.

SIZE in inches.	75 FEET HEAD.			150 FEET HEAD.			225 FEET HEAD.			300 FEET HEAD.		
	Thickness in inches.	WEIGHT		Thickness in inches.	WEIGHT.		Thickness in inches.	WEIGHT.		Thickness in inches.	WEIGHT.	
		Per foot	Per length.		Per foot	Per length.		Per foot.	Per length.		Per foot.	Per length.
2	1/4	6	57	5/16	7½	72	3/8	9	84	7/16	10½	99
3	5/16	11	137	3/8	12½	155	13/32	14	172	15/32	16	198
4	5/16	16	200	3/8	17½	218	13/32	19½	240	15/32	21	260
6	3/8	25	309	13/32	28	343	7/16	30	370	15/32	32½	400
8	7/16	40	490	15/32	43	525	17/32	48	590	9/16	52	640
10	15/32	52	640	1/2	57	700	9/16	63	766	5/8	70	856
12	1/2	67	825	17/32	72	885	19/32	81	1000	11/16	94	1150
14	17/32	84	1036	19/32	94	1165	11/16	107	1335	13/16	125	1535
16	9/16	100	1250	5/8	115	1435	3/4	131	1605	7/8	150	1840
18	19/32	115	1425	21/32	130	1600	13/16	160	1990	15/16	185	2280
20	5/8	135	1665	11/16	153	1895	7/8	190	2340	1	220	2725
24	3/4	195	2425	13/16	212	2610	31/32	250	3090	1 1/16	290	3580
30	25/32	262	3275	29/32	300	3735	1 1/8	365	4565	1 1/4	400	4910
36	7/8	340	4230	1 1/16	413	5125	1 5/16	515	6525	1 9/16	621	7813
42	1	430	5360	1 1/4	545	6780	1 1/2	640	8170	1 5/8	710	9030
48	1 1/16	565	7088	1 3/8	668	8440	1 5/8	845	10540	1 7/8	980	12310

All pipes cast vertically in dry sand in 12 feet lengths, exclusive of bell, except 2-inch, which are 9 feet long. Other weights furnished when required. All pipes are tested by hydraulic pressure to 300 pounds per square inch.

Cast Iron Pipe and Specials.

For Gas or Water Mains.

THE UTICA PIPE FOUNDRY CO. have a large and well equipped Foundry devoted exclusively to the making of Cast Iron Pipe and Specials, for water or gas. We are the sole selling agents for the Company and are at all times ready to quote low prices for large orders, having furnished the entire Pipe and fittings for over one hundred Water Works within the past few years, besides supplying many older companies with all of their extensions. The Pipe is the best made in this country and invariably gives entire satisfaction.

We have received many words of approval from government officials, experienced gas and water works superintendents, eminent engineers and practical contractors who have used our Pipe, but lack of space prevents us from giving them here.

We illustrate a number of the Specials we make on the following pages. We make all sizes, but give the approximate weights of only the smaller sizes to aid in making estimates.

Crosses.

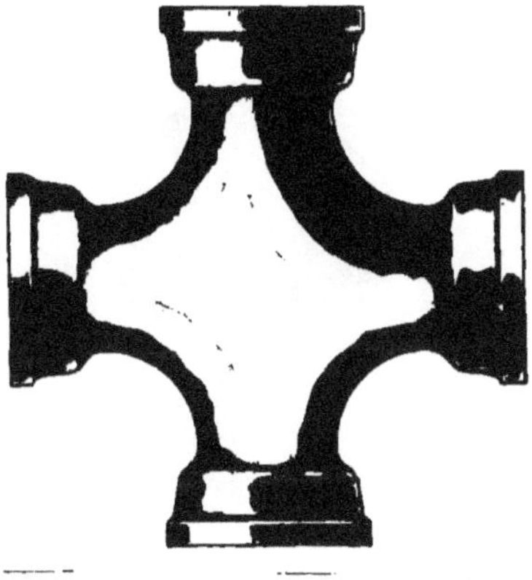

	Lbs.		Lbs.
3x 3x 3x 3 inch	110	8x 8x 8x 8 inch	368
4x 4x 4x 4 "	150	10x10x10x10 "	570
6x 6x 6x 6 "	265	12x12x12x12 "	770

Tees.

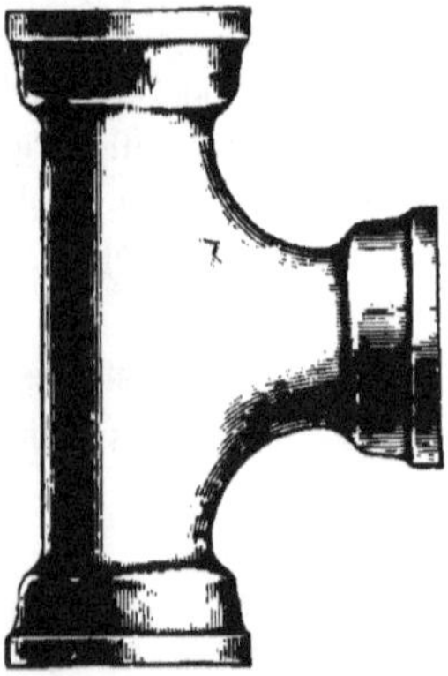

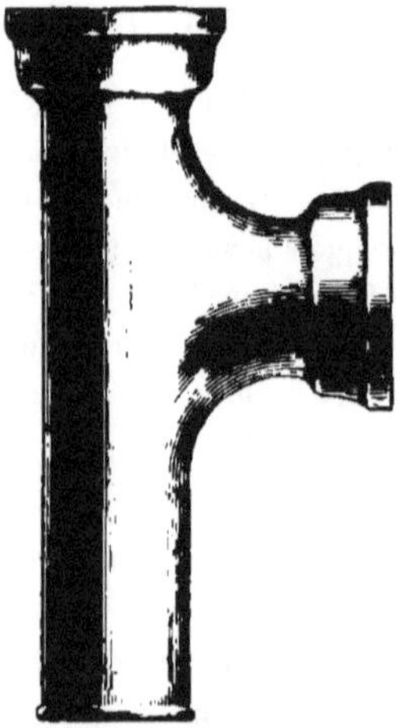

	Lbs.		Lbs.
3x 3x 3 inch	75	8x 8x 8 inch	300
4x 4x 4 "	112	10x10x10 "	452
6x 6x 6 "	210	12x12x12 "	612

Y Branches.

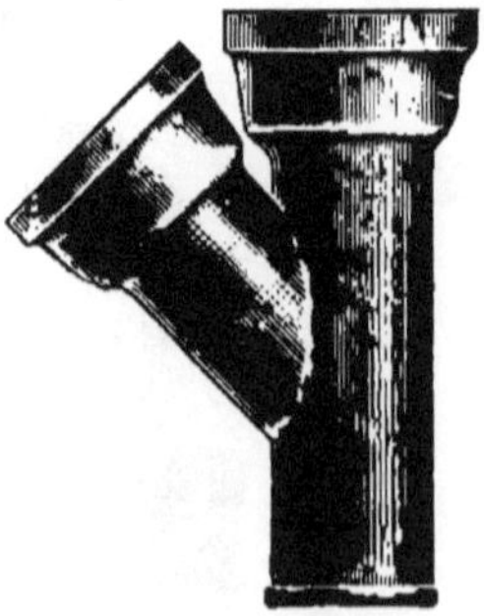

	Lbs.		Lbs.
3x 3x 3 inch	72	8x 8x 8 inch	280
4x 4x 4 "	106	10x10x10 "	430
6x 6x 6 "	206	12x12x12 "	600

Quarter Bends.

		Lbs.
3 inch		50
4	"	71
6	"	136
8	"	200
10	"	248
12	"	348

Eighth Bends.

		Lbs
3 inch		39
4	"	53
6	"	94
8	"	120
10	"	170
12	"	352

Sixteenth Bends.

		Lbs.
3 inch		31
4	"	43
6	"	75
8	"	110
10	"	135
12	"	258

Drip Pot.

Reducers.

	Lbs.
3 x 2 inch	40
4 x 3 "	45
6 x 4 "	80
8 x 6 "	120
10 x 8 "	169
12 x 10 "	303

Reducers.

	Lbs.
3 x 2 inch	35
4 x 3 "	40
6 x 4 "	60
8 x 6 "	98
10 x 8 "	120
12 x 10 "	285

Plugs.

	Lbs.
3 inch	6
4 "	9
6 "	17
8 "	27
10 "	37
12 "	51

Caps.

	Lbs.		Lbs
3 inch	16	8 inch	50
4 "	22	10 "	76
6 "	35	12 "	100

Sleeves.

	Lbs.
3 inch	40
4 "	53
6 "	82
8 "	102
10 "	125
12 "	175

Split Sleeves.

	Lbs.
3 inch	52
4 "	63
6 "	106
8 "	129
10 "	179
12 "	2

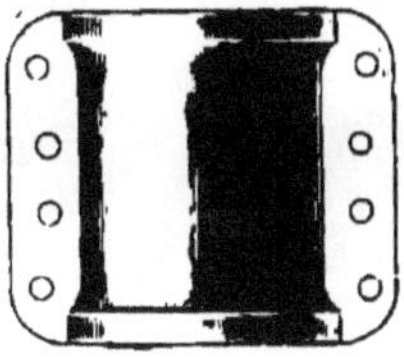

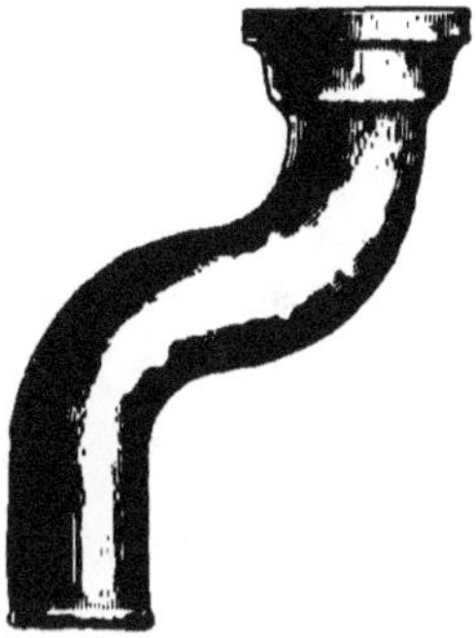

S Bend.

Foot Elbow.

Section of Pipe.

Showing Shape of Bell and Spigot Ends.

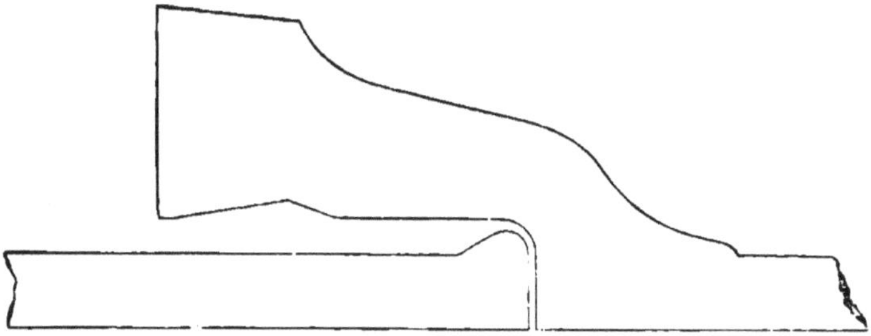

Flanged Fittings.

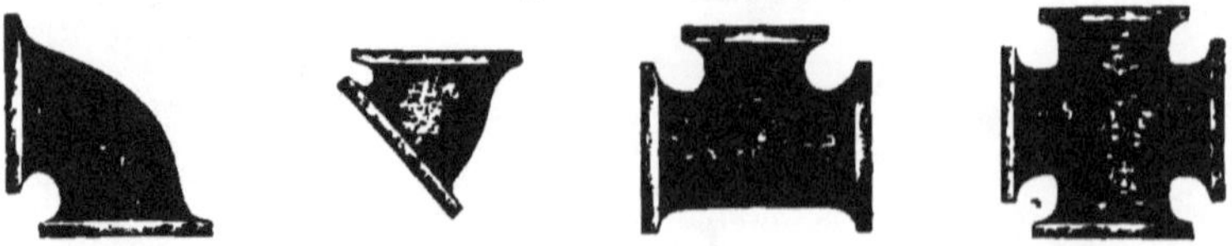

The above cuts illustrate the Standard Styles of Flanged Fittings, which are in growing demand for fitting up large and modern steam plants. These we do not keep in stock, but can furnish to order promptly. Dimensions and prices furnished on application.

Steam Pumps.

We keep constantly on hand a large stock of all the leading and best makes, and can furnish same at manufacturers' prices.

Circulars mailed on application.

The Excelsior Grease and Sewer-Gas Trap.

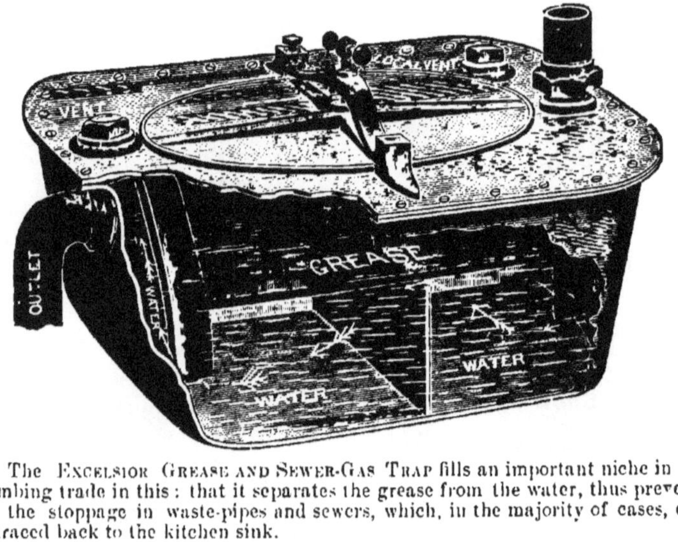

The EXCELSIOR GREASE AND SEWER-GAS TRAP fills an important niche in the plumbing trade in this: that it separates the grease from the water, thus preventing the stoppage in waste-pipes and sewers, which, in the majority of cases, can be traced back to the kitchen sink.

The Trap is designed to be placed in under the sink. Since it is perfectly air-tight, no sewer gas or odor of any kind can escape from it. The top is bedded in cement, and the hand-hole compressed on a flexible rubber gasket. The opening for ventilation of waste can be used, if desired, by removing plug, as also can the opening for local vent.

SIZE, 18 in. long, 12 in. wide, 10 in. deep.

Each..$20.00

www.ingramcontent.com/pod-product-compliance
Lightning Source LLC
Chambersburg PA
CBHW021528210326
41599CB00012B/1425
9783337377076